Sex in an Age of Technological Reproduction

ICSI and *Taboos*

Carl Djerassi

The University of Wisconsin Press

The University of Wisconsin Press
1930 Monroe Street, 3rd Floor
Madison, Wisconsin 53711-2059

www.wisc.edu/wisconsinpress/

3 Henrietta Street
London WC2E 8LU, England

Copyright © 2008
The Board of Regents of the University of Wisconsin System
All rights reserved

5   4   3   2   1

Printed in the United States of America

Library of Congress Cataloging-in-Publication Data
Djerassi, Carl.
   [ICSI]
   Sex in an age of technological reproduction : ICSI and Taboos / Carl Djerassi.
      p. cm.
   ISBN 978-0-299-22790-6 (cloth: alk. paper)
   ISBN 978-0-299-22794-4 (pbk.: alk. paper)
     1. Human reproductive technology—Drama.   2. Medical ethics—Drama.
3. Fertilization in vitro—Drama   4. Artificial insemination, Human—Drama.
I. Djerassi, Carl. Taboos.   II. Title.
   PS3554.J47I27   2008
   812'.54—dc22   2008010993

These plays are fully protected by the author's copyright and any filming, reading, or performance of any kind whatsoever must be cleared beforehand with the author (djerassi@stanford.edu, www.djerassi.com).

With minor modifications, scene 4 of *ICSI* was first published in the Medicine and Creativity special issue of *The Lancet* 368 (Dec. 2006): S55–S57. Reprinted with permission of Elsevier Properties SA.
   The film of an ICSI procedure on the accompanying disc is based on an actual fertilization conducted by Dr. Barry R. Behr of Stanford University.

Contents

Preface vii

ICSI: A Pedagogic Wordplay for Two Voices
   with Audiovisuals 1

Taboos: A Play in Two Acts 35

Preface

Prospective readers who approach a book in the old-fashioned way, by browsing through the pages in a bookstore, may be inclined to close this one rapidly when they notice that virtually everything is presented in direct speech. "It's a *play!*" they might mutter, slamming the covers.

But I am convinced that *reading* an intellectually challenging play, as opposed to solely *seeing* it on the stage, can be incomparably exciting and stimulating. Presenting information in the form of dialogue produces an effect of vivacity and immediacy and permits complex arguments to remain unresolved—as they are in life. Yet since the beginning of the Age of Enlightenment, dialogue has essentially disappeared from the written discourse of scientists.

One of my objectives as a scientist-turned-literary-author is to smuggle scientific issues into the minds of an uninterested or even antagonistic public, by means of the most dialogic form of literature, drama. Of course I wish each of my plays to be seen on the stage. But I believe that all of them can also be read on their own merits as tales of science and scientists, in a format that reproduces direct interaction among human beings rather than in the voice of a third person, "omniscient" narrator. Experimental, albeit anecdotal, confirmation for this assertion is provided by another of my "science-in-theatre" plays, *Oxygen*—written jointly with

Roald Hoffmann—which, within four years, had been translated into ten languages and published in book form in seven of them. *Oxygen* has not only been performed in many countries but it is also widely read, and I hope that the present duet will meet with the same good fortune.

ICSI and *Taboos* are primarily works of "science-in-theatre." By this label I refer to plays in which science or scientists do not just fulfill a metaphoric function, praiseworthy and attractive as such efforts have been in major plays by dramatists such as Brecht, Dürrenmatt, Stoppard, and others. In my plays, what I call the "tribal practices" of scientists constitute the central focus of the drama, as, for instance, in Michael Frayn's *Copenhagen*. My concept of "science-in-theatre" requires that the science depicted be actual or at least plausible and that the conduct of my scientific characters be authentic documentations of professional behavior. These same self-imposed restrictions pertain also to the form of literary prose I have produced during the past eighteen years in a tetralogy of novels that I have categorized generically as "science-in-fiction" to differentiate it from the much more widely practiced genre of science fiction.

That a scientist who turned late in life into a novelist and then into a playwright should dip into his former professional life for inspiration can hardly be surprising. As a chemist who was actively involved in the first synthesis of an oral contraceptive, the subject of human reproduction—or more precisely, the control over reproduction—has interested me for decades. It was inevitable that I should use sex and reproduction, these most personal of personal human experiences, as appropriate themes to explore in my plays. So let me start with some general comments about reproduction before addressing the specific themes developed in the present volume.

With very few exceptions, the millions of different species on this earth—from insects and reptiles to fish, birds, and mammals—copulate in order to procreate. Generally, such procreation carries no personal stamp of knowledge: what is created in that act is not primarily an extension of the individual but rather a preservation of the species. With few exceptions, most males other than humans do not actually know who their offspring are, nor do the fathers of most species have anything to do with rearing the next generation.

Not so with man. Parenthood generally elicits a deep, personal association with one's children, which is often expressed by obsessive identification with them. It takes little imagination to relate the desire for parenthood to a desire for a form of immortality, even at such simple a level as perpetuation of one's family name. Once we recognize this obsession, many of the traditional attempts at regulating sexuality take on a new significance. Until recently, becoming a biological parent invariably meant achieving successful fertilization of a woman's egg by a man's sperm through heterosexual intercourse. Many religions, Catholicism being a prime example, insist that sexual intercourse not only be monogamous, thus clearly defining the biological identity of the offspring, but also that it be sanctioned only if reproduction is its conscious objective. Judaism, on the other hand, through its conferral of Jewish identity through the mother rather than father, tacitly acknowledges the uncertainty of paternal credentials—itself an increasingly outdated notion in this age of DNA testing. But these attempts at confirming the identity of the offspring are not all that seems to govern our traditional sexual mores: it does not so neatly explain, for instance, the Catholic Church's disapproval of contraception, which seems at times reducible to the injunction, "You cannot have sex just for fun."

In *Taboos* I assign such religious dictates to a Fundamentalist Christian couple, whose beliefs about "proper" sexual conduct are often very strict. For example, in 2005 an American radio program titled *New Life Ministries,* which broadcasts on 150 stations nationally and promotes Bible-based abstinence from pornography, adultery, nonmarital sex, and masturbation, summarized its position in starkly unambiguous language reminiscent of that employed by my play's characters: "Our goal is sexual purity. You are sexually pure when no sexual gratification comes from anyone or anything but your wife."

But must reproduction always be initiated by sexual intercourse? Two of the most startling developments in contemporary science, stimulants of enormous social change, have radically disrupted the historically unquestioned connection between sex and reproduction. The first of these was the invention in 1977 of IVF (in vitro fertilization) by Edwards and Steptoe in the U.K. The second—an even more remarkable advance—has been the use of an assisted reproductive technique called ICSI (intracytoplasmic sperm injection) developed by Belgian scientists in 1991.

It is the generation now in school that will reap the consequences of the invention of ICSI that is revolutionizing the nature of human reproduction, pushing to the limits the consequences of the impending separation of sex ("in bed") and fertilization ("under the microscope"). They and *their* children will be the first two generations in which even *fertile* couples will start to use techniques of assisted reproduction for having children. It is essential that these new generations understand the basis of such techniques and the enormous consequences—bad or dubious, as well as beneficial—that such practices may engender. The purpose of my writings on the subject is not to provide answers but to raise questions, because reproductive decisions are made in privacy, within the dynamics of a relationship and the cultural milieu of the couple.

To explain the scientific basis of the ICSI technique in as painless and engaging a way as possible, *ICSI* is presented in the first half of *Sex in an Age of Technological Reproduction* in the form of a script of a simulated TV interview—a type of "verbal combat"—between two persons. Illustrative visuals are contained on the accompanying DVD. The play is intended to stimulate active debate regarding the ethical issues associated with the birth of children conceived without sexual intercourse. Abnormal as that may sound, over the past thirty years there have already been more than two million persons born via assisted reproductive technologies rather than through coitus by parents with impaired fertility. What will change in the future is the extension of this technique to fertile couples. Its greatest beneficiaries may be women who decide to postpone childbearing until their late thirties or early forties because they do not wish to have to choose, in their twenties, between motherhood and professional careers.

That is why I've labeled *ICSI* a "pedagogic" work. I envisioned its main use as a classroom text, with students playing the interview roles; indeed, *ICSI* has already been performed in just this way in many classrooms in Germany, Italy, and Taiwan. But it has also been presented at some professional meetings, medical congresses, and graduate programs, as well as in a performance on BBC Radio 3, proving that this type of dialogic format can also be employed effectively outside the conventional classroom.

No doubt, many readers of the present book will be unfamiliar with the term ICSI. Yet I am confident that, once they have seen the injection of a single sperm into an egg in the accompanying video, they will understand the ICSI technology and will never forget it or its ethical ramifications. If so, such "science-in-theatre"—whether seen on the stage, performed in the classroom, or read at home—will have bridged, however briefly, the widening gulf between the natural and social sciences. *Taboos* then follows naturally by emphasizing the socio-cultural consequences of assisted reproduction, which in many respects are even more complicated than the underlying technology.

ICSI

A Pedagogic Wordplay for Two Voices with Audiovisuals

Definition

Impregnation of a woman's egg by a fertile man in normal intercourse requires tens of millions of sperm—as many as one hundred million in one ejaculate. Successful fertilization with one single sperm is a total impossibility, considering that a man ejaculating even one to three million sperm is functionally infertile. But in 1992 Gianpiero Palermo, Hubert Joris, Paul Devroey, and André C. Van Steirteghem from the University of Brussels published their sensational paper in *The Lancet* (vol. 340, pp. 17-18), in which they announced the successful fertilization of a human egg with a *single* sperm by direct injection under the microscope, followed by reinsertion of the fertilized egg into the woman's uterus. ICSI—the accepted acronym for intracytoplasmic sperm injection—has now become the most powerful tool for the treatment of male infertility: well over 150,000 ICSI babies have already been born since 1992.

This is the factual background of *ICSI*. But because this pedagogic experiment is presented as a wordplay, the characters, though not the actual science, are fictional—especially Dr. Melanie Laidlaw, ICSI's putative inventor. The ethical problems raised by ICSI, however, will linger long after the last word has been spoken in this "interview."

Cast

DR. FELIX FRANKENTHALER   middle-aged American clinician and infertility specialist.

ISABEL YOUNGBLOOD   host of TV program *Dissection;* early to middle thirties, smart and not too subtle critic of science and technology; stylishly dressed, preferably in pantsuit.

Time

Friday, the thirteenth.

Location

TV studio of weekly "issues" program entitled *Dissection;* equipped with two comfortable chairs, perhaps low coffee table, and suitable screen for projection of images.

Technical Details

The "TV Program" scenes require the occasional brief display of visual images while scene 3 incorporates a brief video film of an actual ICSI injection, both of which can be found on the accompanying DVD.

Pedagogic Function

This play is written for classroom use in lieu of a conventional fifty-minute lecture and is envisaged as a staged reading by two persons (preferably students) using audiovisuals. It was tested in this manner before several thousand high school students in Germany and hundreds of college students in the United States. A key pedagogic component is the subsequent discussion dealing with the numerous practical and ethical issues raised through such a "theatrical" presentation. Ideally, each student in the audience is provided with a copy of this text that can then be reread at home together with the material included on the DVD, with the discussion and other feedback then occurring during a second class period. However, alternative modes can also be used, such as spreading the scenes over several classes with discussions following each presentation or using a single abbreviated version (as indicated in the text) so as to include discussion within the entire fifty-minute time frame. It should be noted that the wordplay has also been presented in Europe to professional audiences at several medical congresses with equally engaged post-performance discussions.

Scene 1

(*Friday, the thirteenth. TV studio. Ideally, the backdrop bears a large sign, "Dissection," followed by a second line, "With Isabel Youngblood."* YOUNGBLOOD *is lounging comfortably in one of the two side chairs, examining a page with notes.* FRANKENTHALER *enters.*)

YOUNGBLOOD (*rises from chair*): Dr. Frankenthaler! Welcome to *Dissection*. I'm your host, Isabel Youngblood

FRANKENTHALER (*stretches out hand*): Thanks for inviting me.

YOUNGBLOOD (*points to other chair*): Please make yourself comfortable. Let's talk a little bit before we go on the air. (*Looks at watch.*) Couple of minutes . . . no more. We want to keep it fresh and unrehearsed. I presume you've watched our program, so you must know the format—

FRANKENTHALER (*slightly uncomfortable*): I'm afraid I'm not a regular watcher—

YOUNGBLOOD: Which ones have you seen?

FRANKENTHALER (*more uncomfortable*): Only the one about minks.

YOUNGBLOOD: Last week's? Too bad you haven't watched more of them to get a flavor of the range of our topics. (*Shrugs dismissively.*) Still, you know how we do it: one guest only . . . yet covering a lot of bases. . . . You know . . . different perspectives. For instance, more of a woman's spin—

FRANKENTHALER: I suppose when the subject allows it—

YOUNGBLOOD: Which it usually does. (*Grins.*) But even if it doesn't, we dissect until we see what's underneath.

FRANKENTHALER (*uncomfortable*): A question. May I?

YOUNGBLOOD (*looks at watch*): Of course.

FRANKENTHALER: Your program about minks—

YOUNGBLOOD (*interrupts somewhat defensively*): What about them? Don't you like minks?

FRANKENTHALER: No, no! Minks are fine. My question was about the focus of your program. I'm afraid, I didn't see all of it. What exactly was it about? Mink farming? The fur industry? Cruelty to animals? Waste disposal?

YOUNGBLOOD: All of the above. (*Again grins.*) And then some. Actually it's good you saw so little. We prefer to have guests without prior prejudices. So let's forget about minks and get to tonight's program. After all, you are here for ICSI and not minks.

FRANKENTHALER: You're right. (*Firmer tone.*) Let's take ICSI. You know I've provided your program with some audiovisuals—

YOUNGBLOOD (*grudgingly*): The producer told me.

FRANKENTHALER: I want to be sure they can be shown at the right moments.

YOUNGBLOOD: Of course they *can* be shown. But are they necessary? We aren't so much interested in the science behind ICSI as—

FRANKENTHALER (*shocked*): But—

YOUNGBLOOD (*holds up hand, then motions with thumb to program title behind her*): This is an *issues* program. We want to dissect the *issues* created by ICSI.

FRANKENTHALER: A point made perfectly clear by your producer.

YOUNGBLOOD: So what's the problem?

FRANKENTHALER: Problem? I have no problem. I just want to be sure your producer informed you that I'd be willing to discuss ICSI *issues,* provided you're prepared to listen first to the science behind ICSI. And for that, I need PowerPoint slides.

YOUNGBLOOD (*conciliatory*): I have no problem with the science. But why slides? Why not just *tell* us—not just me, but our public?

FRANKENTHALER (*curt*): We scientists like pictures. Besides, they save time.

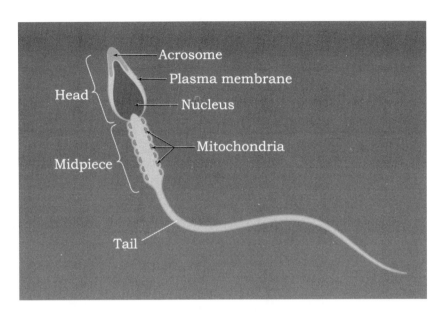

Figure 1

YOUNGBLOOD: Let *me* worry about time on this program.

FRANKENTHALER (*slyly*): Ms. Youngblood. Do you know what an acrosome is?

YOUNGBLOOD: Spell it.

FRANKENTHALER (*spelling it slowly*): A C R O S O M E. So what's an acrosome?

YOUNGBLOOD: You tell me!

FRANKENTHALER (*openly pleased*): Now—before the program starts—could you quickly have them project my first image?

YOUNGBLOOD (*dismissive shrug, then points to invisible cameraman*): Lou! Give us the first one. But make it snappy.

(*Figure 1, showing stylized picture of a sperm, appears on screen.*)

FRANKENTHALER (*quick glance at image, then turns to* YOUNGBLOOD): So what do you see there, Ms. Youngblood?

YOUNGBLOOD (*turns toward screen and remains in that position*): A picture of a sperm, of course.

FRANKENTHALER: Consisting of what?

YOUNGBLOOD (*reads*): Head . . . midpiece . . . tail.

FRANKENTHALER (*ironic*): *Very* good! But that doesn't really tell you very much about their function, does it? For instance, do you know what the midpiece of a sperm does? (*Pause.*) Not very likely. So let's turn to the four words on the right side of the slide. *Miz* Youngblood . . . or is it *Doctor?*

YOUNGBLOOD (*feigns sweet smile*): Plain Miz.

FRANKENTHALER: Now then, Ms. Youngblood. I'm prepared to offer odds that of the four words on the right side, at least two are unfamiliar to much of your audience. I would offer even bigger odds that the two words are "mitochondria" and "acrosome" and that even spelling them won't help. Right?

YOUNGBLOOD (*impatient*): Better make your point! The program is about to start.

FRANKENTHALER: The mitochondria are the engine and the fuel tank that power our sperm. The acrosome, shown here in orange on the very tip of the sperm, houses the explosive—actually a group of enzymes—that will permit penetration of the egg's protective shell. Remember, the sperm has to get *inside* the egg to effect fertilization. If you don't know how *that* happens, ICSI won't make much sense.

(*Pauses, while looking first at* YOUNGBLOOD *and then at sperm image.*)

The term acrosome is unfamiliar to most men, yet it ought to be as much an every-day word as "uterus" is to you and most women. During your program, I hope to be able to show you why. Only then will I participate in your dissection.

(*Again pauses, while looking at* YOUNGBLOOD.)

Is that a deal? I'm ready for your *issues,* if you're ready for the *science.*

YOUNGBLOOD: It's a deal. By the way, you aren't superstitious are you?

FRANKENTHALER (*faint smile*): Few scientists are. Why do you ask?

YOUNGBLOOD: Today is Friday, the thirteenth.

(*End of scene 1.*)

Scene 2

(*TV studio. The backdrop still bears the sign "Dissection," followed by a second line, "With Isabel Youngblood."* YOUNGBLOOD *and*

FRANKENTHALER *are sitting so as to face partly each other as well as the audience. As soon as the actual TV program starts, the sign is extinguished.*)

YOUNGBLOOD (*straightens in chair, looks at camera*): Welcome to *Dissection*. I'm your host, Isabel Youngblood. Tonight, we plan to dissect ICSI—a word unlikely to be familiar to most of you. Yet when we're finished, you are unlikely to ever forget it. To help us understand ICSI, we are fortunate to have with us Dr. Felix Frankenthaler, one of the fathers of ICSI. (*Smiles, while gesturing toward* FRANKENTHALER.) Good of you to join our program.

FRANKENTHALER: My pleasure.

YOUNGBLOOD (*addresses camera*): Just before we went on the air, Dr. Frankenthaler informed me that he was not burdened by superstition. That he didn't mind us dissecting his ICSI baby on a Friday, the thirteenth. (*Pause.*) So what *is* ICSI, Dr. Frankenthaler?

FRANKENTHALER: An acronym standing for (*Slows down to enunciate clearly.*) intracytoplasmic . . . sperm . . . injection, in other words—

YOUNGBLOOD (*quickly interrupts*): One-shot fertilization. (*Mock apology.*) Sorry, I shouldn't have said that! This is a serious topic. How about "Fertilization of an egg under the microscope by injection of a single sperm"? Precise and unambiguous.

FRANKENTHALER (*wags head*): Precise? Yes. But unambiguous? Some viewers may ask, "Doesn't all *natural* fertilization involve the entry of one sperm into an egg?" May I show you how I'd handle that?

YOUNGBLOOD (*ironic, with après vous gesture*): Why not?

FRANKENTHALER: So what's so unusual about ICSI? About the ability to fertilize an egg by the injection of (*Slows down.*) one . . . single . . . sperm? It *is* unusual—indeed unprecedented—because in normal sexual intercourse, tens of millions of sperm are ejaculated to fertilize an egg. A man with only a few million sperm—seemingly still a very large number—is functionally infertile, because overwhelming the egg's defenses requires the sheer force of enormous numbers. A normal man has, say, fifty to one hundred million sperm in a single ejaculate.

YOUNGBLOOD: That reminds me of a joke—a woman's joke: "Why does a man produce so much sperm in one throw?" The answer is, "Because sperm never ask for directions."

ICSI | 9

(*Brief pause, during which audience presumably laughs. If not,* YOUNGBLOOD *should say, "Her lousy joke, not mine."* YOUNGBLOOD *notices* FRANKENTHALER *shaking his head in disapproval.*)

You don't find that funny?

FRANKENTHALER: The gag may be funny, but the subject isn't. Not if you're an infertile man.

YOUNGBLOOD (*conciliatory*): Still, a hundred million is an awesome number. Do men really need to ejaculate so much?

FRANKENTHALER (*nods curtly*): The answer is yes. We do. Because in the female reproductive tract, sperm face terrible odds in their journey to the egg. Let me illustrate that with a picture.

(YOUNGBLOOD *gestures to unseen camera, whereupon Figure 2 is projected.*)

You will note that while in ordinary intercourse, up to one hundred million sperm are deposited in a woman's vagina at the start of the race to the ovum, only a few thousand are left when the surviving sperm finally make it to the waiting egg.

(*Pause while* FRANKENTHALER *points with laser pointer to appropriate spot on the picture: first to spot marked "10 million sperm," then to spot marked "1 million sperm," then to spot marked "100,000 sperm," finally resting somewhat longer on "? sperm" in picture.*)

To illustrate the sperm's problems—in other words the seminal dilemma—let's look quickly at my next image.

(*Turns toward* YOUNGBLOOD, *who in turn gestures to off-stage camera man, whereupon Figure 3 appears on screen.*)

This is an actual micrograph showing a sperm's problem as it struggles through the thick, sticky mucus of the cervix. But only then does the *real* competition begin: the struggle to penetrate the egg itself. As the egg resists with a battery of chemical defenses, only a single sperm can win. All of which is to explain why, if a man's sperm count falls to just a few million sperm, it makes him, for all practical purposes, infertile.

YOUNGBLOOD: I've heard it said that this sperm race (*Points behind her to picture on screen.*) is no Olympic event. That there's only a gold medalist. That silver and bronze medals are not awarded. In other words, that only one sperm can enter the egg. Would you explain to us why that is so?

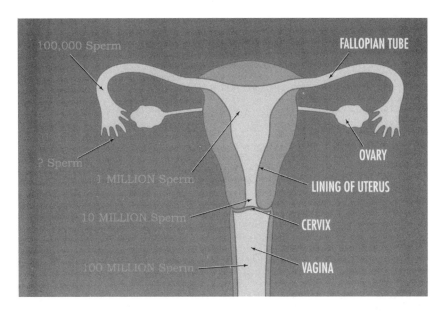

Figure 2

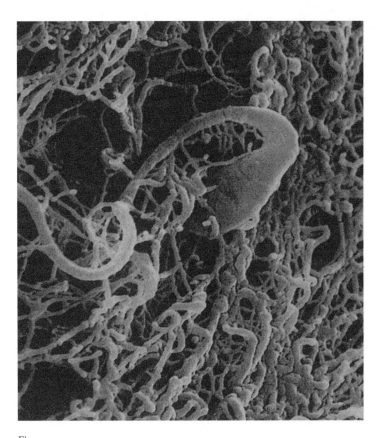

Figure 3

FRANKENTHALER (*pleased*): Gladly. The quickest way to do that—

YOUNGBLOOD (*looks down at her notes*): One moment, Dr. Frankenthaler. Before you do, there is one question about tonight's topic I should've asked right at the beginning.

FRANKENTHALER: Yes?

YOUNGBLOOD (*ready to draw blood*): You should first explain to us why *you* are taking credit for ICSI. I thought a *woman*—Dr. Melanie Laidlaw—was the inventor. In fact, why are *you* here and not Dr. Laidlaw?

FRANKENTHALER (*undisturbed, even amused*): I guess because your producer invited *me*.

YOUNGBLOOD (*annoyed*): You know that's not what I meant. I want to know why you call yourself "a father of ICSI."

FRANKENTHALER: *You* called me that. I've never claimed to be the inventor. But like most babies, ICSI needs two parents. Ironically, Dr. Laidlaw—a woman of our age, but clearly ahead of her time—has focused almost exclusively on the sperm under the microscope, whereas I—whom you called "the father"—dealt with the subsequent transformation, *in the woman's body,* of the fertilized egg into a live baby. And since I gather that you want to apply your (*Sarcastic.*) scalpel to the *issues* arising from ICSI, your producer probably thought that I'd be the better target for you. Come to think of it, perhaps Dr. Laidlaw should be called the father—after all, *she* was the one who injected the sperm into the egg—and I the mother. Satisfied?

YOUNGBLOOD: Let's move on to ICSI.

FRANKENTHALER: Not so fast! I haven't answered your question yet why only one sperm can enter an egg. May I?

(YOUNGBLOOD *nods reluctantly.*)

Let's have the next picture.

(*Figure 4 appears on screen.*)

Here we have the egg surrounded by the relatively few sperm that made it to the final round. The blue ring, the outermost barrier to be penetrated, is called the zona pellucida, which is identified by the arrow on the figure. Let's now focus on the winning sperm. Next picture, please.

(*Figure 5 appears on screen.*)

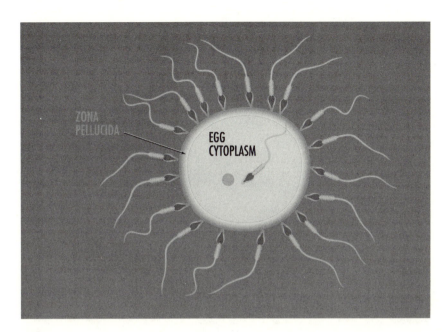

Figure 4

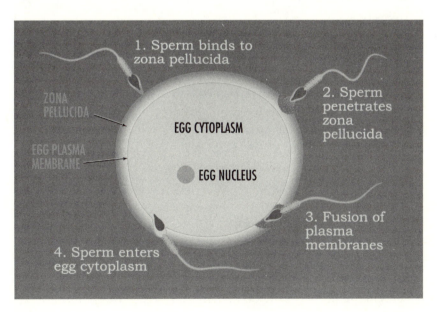

Figure 5

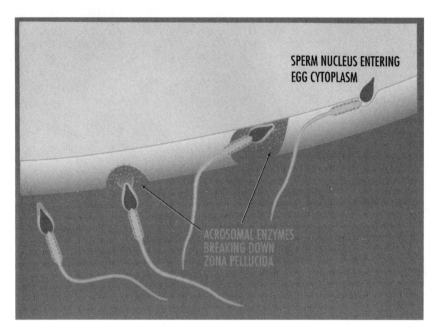

Figure 6

For fertilization to occur, the head of the gold medalist must first bind to the zona, much as a key fits a lock.

(FRANKENTHALER—*with laser pointer—points to legend on screen reading "1. Sperm binds to zona pellucida."*)

Having locked onto its target, the winning sperm now literally bores into the zona pellucida.

(FRANKENTHALER *points briefly on screen to spot reading "2. Sperm penetrates zona pellucida."*)

Now let's look at a close-up of the penetration. (*Raises finger.*) The next figure please.

(*Figure 6 appears on screen.*)

Note how the sperm eats its way in with the help of the orange-colored "explosives" in its warhead, called the acrosomal enzymes. (*Urgent tone.*) And observe how the sperm sneaks in sideways, rather than pushing head on . . . (FRANKENTHALER—*using laser pointer—moves along the screen to the specific spots in Figure 6 in tempo with his voice.*)

until the sperm's nucleus is finally delivered into the interior of the egg. That nucleus, of course, contains the man's DNA—the genetic material—that will eventually fuse with the genetic material of the egg. Only if they fuse have we accomplished fertilization.

YOUNGBLOOD (*who had leaned sideways throughout entire presentation, looking at the screen, now turns to* FRANKENTHALER): But why can't the next sperm waiting outside the zona pellucida repeat that process?

FRANKENTHALER: Good point. In my hurry to get to ICSI, I forgot to mention the key event. The moment that first sperm has sneaked in sideways, it sets off a defensive reaction in the egg membrane—a sort of bomb explosion of its own—that . . . in perhaps oversimplified terms . . . solidifies the zona pellucida's squishy cement into concrete, turning it instantly into an impenetrable barrier.

YOUNGBLOOD (*impressed*): How clever of the egg. (*Pause.*) But if only *one* sperm makes it, how do you get twins? (*Pause.*) Or triplets?

FRANKENTHALER: Another good question. I had not intended to address that problem, because it's a bit complicated, but since you asked—

YOUNGBLOOD: Can you tell us quickly how twins are produced? And then we'll move to ICSI.

FRANKENTHALER: Sure. (*Amused tone.*) But do you mean dizygotic or monozygotic twins? Or both?

YOUNGBLOOD: First, why don't you define those terms for the benefit of our viewers?

FRANKENTHALER: Dizygotic are fraternal twins. We call identical twins monozygotic.

YOUNGBLOOD: Tell us about both. My grandmother was an identical twin.

FRANKENTHALER: Twins and triplets do have something to do with ICSI. But to answer your question, I need to go back to natural fertilization. (*Notices her disappointment.*) I promise to be quick. And then it's ICSI all the way. Okay?

YOUNGBLOOD: I suppose so.

FRANKENTHALER: I will use one more slide to explain the very early steps of embryonic development. Remember? We have just fertilized an egg—be it by natural fertilization through sexual intercourse or through direct injection under a microscope via ICSI. Now look at the next image.

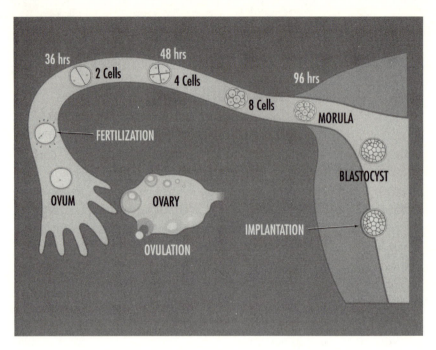

Figure 7

(*Motions to offstage cameraman. Both* YOUNGBLOOD *and* FRANKENTHALER *turn toward screen as soon as Figure 7 appears.*)

Let's assume that we are dealing with ordinary fertilization. The woman has released a single egg from her ovary at the time of ovulation, say day fourteen of her menstrual cycle. (FRANKENTHALER *with laser pointer points to small egg departing from ovary.*) That egg, labeled "ovum" in the picture, has just been fertilized by a single sperm in the manner outlined before. (FRANKENTHALER *points to "fertilization."*) Once fusion of the egg and sperm nuclei has occurred, the first cell division is observed within thirty-six hours. (FRANKENTHALER *points to spot labeled "2 cells."*) Twelve hours later, cell division is repeated (FRANKENTHALER *points to spot labeled "4 cells."*), meaning that embryo formation is on the way as the fertilized egg moves along the fallopian tube toward the uterus. (FRANKENTHALER *points to spot labeled "8* cells.") When it reaches the so-called morula—meaning "mulberry"—stage, we have already sixteen identical cells. And now to your question. (*Looks at* YOUNGBLOOD.) Every once in a while, at this eight- to sixteen-cell stage, separation occurs into two or more identical pieces, which then develop separately like the *undivided* morula shown in the figure. (*He points to spot labeled "morula" and then to "blastocyst."*) If such rare separation occurs, you end up with identical twins. You follow me?

YOUNGBLOOD: Of course. But so far, it's still a one-sperm, one-egg proposition.

FRANKENTHALER: Correct. And that's why they are *identical* twins. Now, going once more to the beginning of this picture . . . every once in a while, the ovary releases two, rather than just one egg . . . (FRANKENTHALER *points again to single egg departing from ovary*.) in which case *each* egg can be fertilized by a *separate* sperm. If that occurs, you get fraternal, or dizygotic, twins.

YOUNGBLOOD: And now to ICSI.

FRANKENTHALER (*nods*): In ICSI, the fertilization (FRANKENTHALER *with laser pointer points to "fertilization."*) is carried out under the microscope by injection of a single sperm into an egg flushed out of the woman. The egg is then incubated for about seventy-two hours (FRANKENTHALER *with laser pointer points to spot marked "8 cells."*) to make sure that embryo formation has started in the normal way before transferring that early-stage embryo back into the woman's uterus.

YOUNGBLOOD: And that's it for ICSI?

FRANKENTHALER: That's it. After that, nature again has to take its course. The embryo, now called "blastocyst," has to implant in the wall of the uterus (*He points to spot marked "blastocyst" and then "implantation."*) and develop for some eight-plus months to yield a baby.

YOUNGBLOOD: Can we now focus on ICSI?

FRANKENTHALER: Shoot.

YOUNGBLOOD: We don't use guns on this program. Only scalpels. (*Pause.*) Now, let's look at the brief film you brought us.

(*End of scene 2.*)

Scene 3

>(*TV studio. The one-minute film of an actual ICSI fertilization has just been shown—perhaps with following ad-libbed commentary by* FRANKENTHALER *using laser pointer at appropriate stages.*)

FRANKENTHALER (*optional commentary*): You are looking at some rapidly moving sperm. But in order to get one into the capillary, which is only one tenth the thickness of a human hair, a sperm has to be immobilized and that is accomplished by crushing its tail with the capillary. Using a bit of suction, the sperm is then aspirated tail first into the capillary. (*Pause.*) You are now seeing an egg, arranged in such a way that the polar body . . . the little "head" on the top . . . is situated in the 12 o'clock position. While the egg is held firmly in place by means of a suction pipette, the egg membrane is now penetrated with the tip of the injection pipette from the 3 o'clock position, the sperm then expelled by pressure and the pipette carefully withdrawn, leaving the single sperm behind. Note that the cell membrane is barely affected and heals within a few minutes.

>(*During that time,* YOUNGBLOOD *and* FRANKENTHALER *turned their chairs partly toward screen. The image of the egg fertilized by ICSI is still projected on the screen as they rearrange their chairs to again face the audience.*)

YOUNGBLOOD (*looks directly into camera*): That was a video of the very first ICSI fertilization in history performed in 1991 by Dr. Melanie Laidlaw, whom Dr. Frankenthaler (*Points to* FRANKENTHALER.) earlier on called "the father" rather than "the mother" of ICSI. So why did we select this ultimate technical fix to human infertility . . . this scientific triumph . . . as a suitable topic for *Dissection?* Because there are issues with ICSI, with the applications of this powerful technique for assisted reproduction, which go beyond science and technology. Issues that should be debated by members of society at large. So let's jump right into the issue at hand. (*Turns to* FRANKENTHALER *with excessive politeness.*) Dr. Frankenthaler, would you summarize the clinical experience of your fertility center since that first genie of reproductive wizardry escaped from the bottle?

FRANKENTHALER: I'd be happy to. That first ICSI fertilization of a human egg occurred in the dim past, on July 7, 1991. (*Emphasis on "1."*) But a lot of H_2O has flowed under that bridge since that time. I will tell you just how much with some data solely from *my* clinic. During the first four years, we have released 943 of what you just called "ICSI genies" (*Draws quotation marks in the air while looking at* YOUNGBLOOD.) in our fertility center. (*Leans forward in direction of unseen camera.*) There were 382 babies . . . (*Pauses for emphasis.*) yes, you heard right . . . 382 babies! In other words, an overall success rate of close to 40 percent, quite spectacular for such severe cases of male infertility associated with too low a sperm count . . . a condition we in the fertility business call "oligospermia." But there is one point that both the audience and my host here should be aware of. As in all instances of assisted reproduction by in vitro fertilization techniques, there was a high incidence of multiple pregnancies: 128 twins and 34 triplets. (*Turns to* YOUNGBLOOD.) Now you see the relevance of my earlier answer to your question about twins.

YOUNGBLOOD (*sharply*): I see no relevance at all. You're having multiple pregnancies because you transfer more than one embryo—I believe at least three—into the woman's uterus to increase the odds. No wonder you get so many twins and triplets.

FRANKENTHALER: Fair enough. Bear in mind, however, that many of our patients are older couples. And that women over forty suffer pregnancy loss six times more often than women below thirty-five. That's why it is important to increase, as you say, the odds. Still, every once in a while, all of them implant—we hit the jackpot, so to speak. So we counsel the couple—especially in the case of triplets, with all the associated risks to the mother and the potential babies—to consider selective reduction—

YOUNGBLOOD: Why not call a spade a spade?

FRANKENTHALER (*quick and defensive*): Because we're not playing cards, nor do we wish to gamble.

YOUNGBLOOD: You know what I'm talking about! Why not say "abortion" rather than "reduction"?

FRANKENTHALER (*irritated tone*): Because we are in the business of *creating* life, not *terminating* it! Selective reduction means *increasing* the chances for completing a pregnancy.

(FRANKENTHALER *is clearly bothered by introduction of abortion issue. Shakes his head for some seconds.*)

With all of our accumulated experience, we now encounter such a high success rate that we find ourselves with more embryos than we can reasonably re-implant into the mothers. During those first four years, just in our clinic, we accumulated 237 supernumerary embryos, which we cryopreserved.

YOUNGBLOOD: Supernumerary?

FRANKENTHALER (*jaded*): I assume you're not requesting a dictionary definition?

YOUNGBLOOD: Not at all. Just questioning your use of terms. Isn't "supernumerary" just another way of saying "rubbish"? Aren't you coming awfully close to playing God? Deciding who gets to live?

FRANKENTHALER: Since you seem to be so concerned about accurate terminology, let me point out that *supernumerary* is not judgmental. It's just a fancy word for "excess." You must allow me at least some big words. Otherwise, how can I impress you? That's why we *cryopreserved* the excess embryos—"froze them," in ordinary English—rather than discarding them. Satisfied?

YOUNGBLOOD: Not really. You're just begging the question, aren't you? What is the ultimate fate of these (*Purposely uses a precious, possibly sarcastic manner of enunciation.*) cryopreserved, supernumerary embryos?

FRANKENTHALER (*in perceptibly lower tone*): That decision has not really been made . . . certainly not to everyone's satisfaction. It's not really a medical decision, is it? First, to whom do the embryos belong? Once the mother has given birth to the desired number of babies by ICSI, she may not wish to have more implanted into her. Should they then be used by other women? Could they be sold by the egg donor and purchased by an infertile woman—even a postmenopausal one—for transfer into her uterus so that she becomes a quasi-biological but not genetic mother? (*Becomes progressively more agitated.*) And if not, what then? Do we simply keep them indefinitely? Use them in research . . . for instance in the hottest of all fields, stem cell research? Or are they junk and turn to garbage when they are discarded? Who decides what to do with that ever-increasing number of cryopreserved embryos?

YOUNGBLOOD: Precisely. So who does?

FRANKENTHALER: I'm the one who posed the question. It's for you (*Turns to the unseen camera.*) all of you, to answer. . . . But only after you first agree whether a three-day-old embryo . . . consisting of a few undifferentiated cells . . . in a Petri dish is life.

YOUNGBLOOD: It's *potential* life!

FRANKENTHALER: Potential? Yes. But so is an unfertilized egg . . . or a sperm.

YOUNGBLOOD: I'm sure some people will agree with that statement.

FRANKENTHALER: In that case, will you call male masturbation "potential baby killing"?

YOUNGBLOOD: I have a feeling we better move on.

FRANKENTHALER: Before we do, let me make just one point, which I would have expected you . . . a woman . . . to have made. If an egg is injected with an otherwise infertile sperm under a microscope and then put in a Petri dish until cell division is confirmed, are you suggesting that the woman is now pregnant? Or that life has now begun? The egg has to be reintroduced into *her,* into *her* body . . . and it must implant in *her* uterus. Only then can we discuss the question of life. (*Pause.*) Fertilization and pregnancy aren't synonymous.

YOUNGBLOOD: I doubt whether everyone will agree with you.

FRANKENTHALER: Which is why I believe that the decision about what to do with unused embryos ought to be the decision of the individual woman, whose egg was converted into an eight- or sixteen-cell embryo before being cryopreserved. But let's return to ICSI and science.

YOUNGBLOOD: All right, let's. (*Shuffles through her notes.*) We've covered ICSI fertilization (*Pause.*), the subsequent steps of embryo formation (*Pause.*), and reinsertion of the egg back into the woman's uterus. (*Looks at* FRANKENTHALER, *who nods in agreement, and then at camera.*) But there's one more thing. If, as you said, fusion is successful, the road to new life is open. But what form of life? Male or female? (*Looks at* FRANKENTHALER.) Why don't you tell us a bit about how that's decided.

FRANKENTHALER: Now that's easy. Biological gender is determined by a pair of sex chromosomes: females have two X chromosomes, whereas males have one X and one Y. (FRANKENTHALER'*s tone turns emphatic.*)

The sex of the offspring will *always* be controlled by the sperm—*never* by the egg. If the sperm contains a sex-determining chromosome called X, the child will be a girl; if it contains a sex-chromosome called Y, a son is born.

YOUNGBLOOD (*nods*): Right. So men who have complained for millennia that their wives didn't provide them with sons have no one to blame but themselves. Trading wives won't do it. It will always be the roll of the seminal dice.

FRANKENTHALER: Well put! "Seminal dice." (*Chuckles.*) And never loaded. If Henry VIII had viewed tonight's show, some of his wives might have survived.

YOUNGBLOOD: I read somewhere that it's now possible to separate X- and Y-containing sperm. I don't know by what method—

FRANKENTHALER: Flow cytometry.

YOUNGBLOOD: Whatever. My question is: if that separation is now possible, could one select a Y-sperm and use it with ICSI to guarantee a son? In other word, *load* the seminal dice?

FRANKENTHALER: I see what you're driving at. It's trivial. With ICSI, the answer is a definite yes.

YOUNGBLOOD (*suddenly agitated, leans forward, facing* FRANKENTHALER): You say *trivial?* You call guaranteed sex predetermination *trivial?* And what then? Preponderance of male children? Overwhelming preponderance? Will that lead to legalized prostitution or polyandry or more wars or—

FRANKENTHALER: Stop! Is it fair to lay all that at *my* feet? *Any* scientist's feet? Would you want such judgment calls made by a mere technician? (*Indicates, with ironic modesty, that by "technician" he means himself.*) When I used the word "trivial," I meant the technical aspects. I meant, yes, now it's easily possible with ICSI to load the dice. I intended no judgment of the social consequences—none at all. I certainly do not recommend establishing reproductive casinos full of ICSI-loaded seminal dice. Such uses—and hence, such questions—do not concern me.

YOUNGBLOOD: But *shouldn't* you worry about the purposes to which your ICSI—

FRANKENTHALER (*mildly*): *Our* ICSI—not *mine*. We both know that a woman scientist played the seminal role.
   (FRANKENTHALER *grins broadly.*)
YOUNGBLOOD (*stiffly*): I repeat the question. Shouldn't you *scientists* worry about the ethical problems raised by ICSI? *Before* the genie escapes from the bottle?
FRANKENTHALER (*sardonic*): Do I hear you correctly? "*Before* it escapes?" It's out, and there's no way of putting it back in. Fifteen years have passed since the first ICSI genie was released. Now there are well over 150,000 ICSI babies all over the world! All of us, including you, will have to learn to live with ICSI.
YOUNGBLOOD: So you agree there is a problem with ICSI when using purposely separated X- or Y-chromosome containing sperm?
FRANKENTHALER: Social problems? Yes. Especially if it were practiced on a wide scale in countries such as China or India that favor male children. But scientific . . . I mean technical ones? (*Shakes head.*) No, I don't see any there.
YOUNGBLOOD (*examines notes in her lap*): Let's move to an issue with ICSI where there may also be technical problems, shall we? Is that okay with you?
FRANKENTHALER (*wary*): What have you got in mind?
YOUNGBLOOD: Initially, you set out to develop ICSI—
FRANKENTHALER: *We* set out. Not *I* alone.
YOUNGBLOOD: I stand corrected. Dr. Melanie Laidlaw and you—
FRANKENTHALER: And our colleagues.
YOUNGBLOOD (*getting impatient*): All right: You two and your colleagues set out to promote ICSI as a treatment for male infertility caused by an insufficient number of sperm.
FRANKENTHALER: True enough, although *promote* is not the most felicitous word to describe what we did. Still, go ahead.
YOUNGBLOOD: But lately, haven't you branched out to treating infertility from other causes, in some of which the sperm isn't even mature? (*Again looks at her notes.*) In fact, extreme cases such as men, who lack a vas deferens?
FRANKENTHALER: Quite correct.

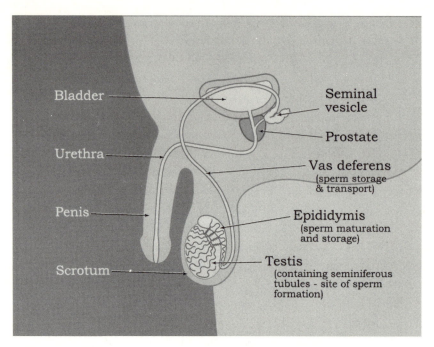

Figure 8

YOUNGBLOOD: I think I will take advantage of the marvels of the television medium to show our audience a schematic picture of the male reproductive organs. (*Smiles.*) And may I reassure our viewers that nothing prurient will flash before their eyes? So now, may I have the picture? *Our* picture, not one of Dr. Frankenthaler's slides.

(*Motions to unseen cameraman and at the same time turns around briefly to check whether correct image has appeared on screen. Turns back to face the camera after seeing Figure 8 on the screen.*)

The vas deferens is the duct in which the sperm is stored and transported. (YOUNGBLOOD *uses a laser pointer to identify vas deferens on the screen image.*) One can well imagine that men born without a vas deferens have not the faintest chance of becoming fathers. Or let's take men who have problems in their epididymis where the sperm matures. (YOUNGBLOOD, *using a laser pointer, marks epididymis on screen.*) That is to say, men whose sperm had not yet acquired the acrosome explosive we have already learned about from Dr. Frankenthaler. He and his colleagues . . . or perhaps Dr. Laidlaw . . . thought that such severely impaired men . . . I mean *reproductively* impaired men . . . are also entitled to ask whether such new reproductive technologies can make them fertile.

(*Pauses for a few seconds while looking at* FRANKENTHALER *and image on screen is turned off.*)

Is that a fair representation of the facts?

FRANKENTHALER: It is.

YOUNGBLOOD (*seemingly innocently*): Would you care telling our audience about the results?

FRANKENTHALER: Gladly. As you stated, we felt that such men need not necessarily be excluded from biological fatherhood, so we tried the ICSI procedure with a few volunteer couples. (*Pauses while slowly looking into the unseen audience.*) It worked . . . like a charm! Once these defective sperm were injected by the ICSI procedure into the egg, everything went according to plan!

YOUNGBLOOD: May I make a brief comment?

FRANKENTHALER (*waves hand magnanimously*): By all means . . . it's your program.

YOUNGBLOOD: Aren't you now operating on the very edge of permissibility? Or even beyond it?

FRANKENTHALER (*frowns, but attempts levity*): Some people will tell you that the only way to find an edge is to fall over it.

YOUNGBLOOD: Are you implying that you have already fallen over it?

FRANKENTHALER: Nothing of the sort. (*Attempts forced chuckle.*) I don't even know how close we are to the edge.

YOUNGBLOOD: Does that mean you will persist until you come crashing down?

FRANKENTHALER: Sometimes, a neutral observer (*Gestures toward her.*) is the first to notice how close—

YOUNGBLOOD: Well, if you're suggesting that people like me should raise the first warning, let me accept your invitation.

FRANKENTHALER: Invitation? I didn't—

YOUNGBLOOD (*raises hand to stop him*): Please! Dr. Frankenthaler, you just mentioned your ability to offer genetic fatherhood to men who are born without a vas deferens. Right?

FRANKENTHALER: Right.

YOUNGBLOOD (*reads from her notes*): Isn't it a fact that such absence of the vas deferens is considered an indicator of cystic fibrosis? (*Her voice acquires an edge of sharpness.*) And that such men, who, of course, are ordinarily infertile, could now have children via ICSI and thus run a significant risk of passing cystic fibrosis on to the offspring? Inheriting the uninheritable? Doesn't that worry you, Dr. Frankenthaler? It worries me. It worries me a great deal.

FRANKENTHALER (*lets out audible sigh*): Of course, you're right about the cystic fibrosis risk and the couple is so advised. We took such an eventuality into consideration at the very outset. We insist on genetic screening of both partners and on pre-implantation genetic screening of the embryo as well as later screening of the fetus. If we find an extra chromosome, indicating mutation, we urge termination of the pregnancy. For your information, so far we have not found any higher incidence of mutations than in ordinary pregnancies. Now, under these conditions, would you not offer ICSI to a man with congenital bilateral absence of the vas deferens?

YOUNGBLOOD (*wags head*): You're still challenging some biological dogmas that were considered inviolable. More than any previous technique, ICSI bypasses several steps that presumably serve as nature's screening mechanisms for deficient sperm. Jumping within one generation over evolutionary barriers that took millennia to become established? You skip the acrosome reaction. You eliminate independent penetration of the zona pellucida and of the egg-membrane. And now you tell us that you can even bypass sperm maturation in the epididymis. Given that one human generation is equal to about twenty-five years, it will take at least that long before the genetic effects of ICSI can be fully evaluated. (*Again wags head.*) By the way, quite separate from the issue of cystic fibrosis, what if the father's infertility is passed on to the son? Are you willing to perpetuate that problem?

FRANKENTHALER: If that's the only thing worrying you, you can relax. All we're perpetuating is a different form of fertilization. If the son is sterile, you'll simply use ICSI again. What's wrong with that? Why are people so preoccupied with non-coital methods of reproduction? (*Becomes progressively more irritated.*) Why always romanticize the act of conception? Why must we dress it up as (*Dismissive.*) "a mating dance between sperm and egg"? And why get upset that ICSI has made that dance superfluous? If you want to dance, dance. If you want to procreate, procreate. Why use procreation as a justification for coitus or dancing? In the final analysis, we are only arguing about differences in delivery vehicles: penis versus pipette. And if anyone wants to pursue this point, a pipette is just a pipette, reproducible to a fraction of a millimeter, whereas penises—

YOUNGBLOOD (*sharply*): Thank you! That's enough!

FRANKENTHALER (*chastened*): You're right. I was carried away. Of course, some people feel that there are aspects of human life that should be off limits to science. Cloning seems to be one. Maybe they'll start adding ICSI to this list if someone starts constructing some horror scenario. At least yours (*Gestures to* YOUNGBLOOD.) doesn't fall into that category. At least not in my book.

YOUNGBLOOD: That's a matter of opinion. (*Pause.*) But if you want horror scenarios—

FRANKENTHALER (*quickly interrupts*): *I* don't want any.

YOUNGBLOOD: Well . . . I do. (*Stares at* FRANKENTHALER *for several seconds before speaking very slowly.*) Tell me. Could ICSI be used with sperm taken from a *dead* man?

FRANKENTHALER (*takes audible deep breath, then sotto voce*): I knew we'd get to this. (*Louder.*) If you're referring to sperm from a dead *fertile* man, the answer is yes. I would say, categorically yes, if the sperm is aspirated within a few hours postmortem, maybe even after twenty-four or thirty hours. It was first done with mice dead twenty-four hours. Indeed, motile sperm was found in antelopes up to *four days* after death, provided the carcass was kept cold. So why shouldn't it work with humans, we asked ourselves? Just so the sperm still shows some signs of viability, say a bit of twitching. Once aspirated, such semen could be preserved for months, if not *years.* (*Seems intrigued by idea and decides to pursue it.*) Perhaps even decades or centuries. Who knows? Kazufumi

Goto, a Japanese animal physiologist, once launched an expedition into the Siberian tundra. He was looking for frozen prehistoric relatives of elephants. To see whether any sperm cells survived that could be injected by ICSI into the eggs of a modern elephant to produce—

YOUNGBLOOD: And you think that's OK?

FRANKENTHALER: You mean experiment with the sperm of woolly mammoths? Why not? They were only a kind of elephant, after all. This is not some *Jurassic Park,* some movie or science fiction, in which the science was fundamentally flawed.

(FRANKENTHALER *makes dismissive gesture.*)

YOUNGBLOOD (*agitated*): Forget the elephants. I'm talking about *us*. About humans. About the next few years . . . not about centuries.

FRANKENTHALER: You asked whether ICSI fertilization with the sperm of a recently deceased man was possible, and I said, yes. It's already been done a number of times with ICSI and so far all the children born from such dead fathers have been normal. You didn't ask whether it was OK.

YOUNGBLOOD: I'm asking now. Is that OK?

FRANKENTHALER (*increasingly irritated, finally losing his cool*): Frankly, I haven't given it any thought and I don't plan to. I can just see where that line of inquiry will lead us. Next, you'll ask what if someone had put Elvis Presley's body into a deep freeze! What if someone bribes the undertaker? What if . . . ? You're trying to push me into territory that I have no intention of traversing. (*Barely controlled anger.*) Do you know what the fundamental difference is between you and me? (*Points to* YOUNGBLOOD.) It is that in your opinion (*Slows down markedly and enunciates each word carefully.*) *if there is a doubt, there is no doubt.*

(FRANKENTHALER *stops and looks challengingly at her.*)

YOUNGBLOOD: I don't get it.

FRANKENTHALER: I didn't think you would. With you, if there's a doubt—any doubt—there's no doubt . . . (*Pregnant pause.*) . . . *that it shouldn't be done*. Whatever "it" is. In this instance, ICSI. You are not prepared to take any risk. With you and your *Dissection* program, it's "noses in, but fingers out." You are willing to stick your nose into every possible question, but not to take any responsibility for wondering how some of those questions can even be answered or whether they're at all relevant to the issue at hand. With most scientists, if there's a doubt you push ahead. Mark you, not heedlessly . . . but you push ahead. (*Pause.*) What science does is tell us what *is*. Yet the public always wants from us *predictions,* whereas we scientists are much better at providing *observations*. You persist in setting up one straw man after another and then setting them on fire. And then you want me to participate in these pyromaniacal activities with barely a fire extinguisher? I trust this won't sound pretentious but where I come from—meaning the real world of infertile couples for whom a baby means everything—if they follow your advice . . . if they keep asking "But what if?" . . . "What if? . . . rather than saying, "Why not?" . . . they might as well forget about procreation and turn to perpetual barrenness.

YOUNGBLOOD (*defensive*): What about adoption?

FRANKENTHALER (*nods curtly*): But what if they don't want to adopt? Or cannot adopt?

YOUNGBLOOD: Cannot?

FRANKENTHALER (*sharply*): You heard right: cannot . . . usually for legal reasons. For instance, single parents; in many countries, gay couples; or people whom adoption agencies consider too old. Take a woman around the menopause. With ICSI—if earlier on she had preserved some of her eggs or ovarian tissue—or without ICSI—using someone else's egg or embryo—she might still give birth to a child.

YOUNGBLOOD (*disdainful*): Dr. Frankenthaler, are you promoting post-menopausal pregnancies? Don't you consider that too old?

FRANKENTHALER (*progressively more irritated*): Some women have an early menopause, others a late one. But you keep asking what *I* think. You should inquire what the individual couple thinks.

YOUNGBLOOD: Not society?

FRANKENTHALER: That also. (*Pause.*) But foremost the couple. So I repeat, what if they cannot adopt?

YOUNGBLOOD: I guess that's their tough luck.

FRANKENTHALER (*openly angry*): You see? That's precisely why I'm now so bothered by the direction in which this discussion has gone. What right do *you*, Isabel Youngblood—or for that matter, your TV program— have to judge such gray options in such starkly black and white terms?

In the final analysis, dissection is not the approach to examining issues of reproduction and infertility.

YOUNGBLOOD: And why not? You, the medical scientist, of all persons, objecting to the process of dissection?

FRANKENTHALER: Yes. I, the infertility specialist. Because what dissection lacks is any sense of compassion. And without compassion, I do not extend to you the privilege of examining infertility.

YOUNGBLOOD: Well . . . that's where we differ. In any event, time's up. Let me thank you for suffering through this week's "dissection" and to our viewing public for joining us. We've all been presented with much food for thought.

(*End of scene 3. In case of time constraints within a standard classroom schedule, the wordplay can be terminated at this point.*)

Scene 4

(*TV studio. Personal discussion between* YOUNGBLOOD *and* FRANKENTHALER *after termination of actual TV program.*)

FRANKENTHALER: Before I leave, may I ask you a question, Ms. Youngblood? Do you have children?

YOUNGBLOOD: I'm not accustomed to being interviewed by my guests.

FRANKENTHALER: We're now off camera. (*Pause.*) Anyway, I take it that the answer is "no." Now I won't ask you how old you are, but I'd guess middle thirties.

YOUNGBLOOD: Dr. Frankenthaler, your personal questions have nothing to do with tonight's topic.

FRANKENTHALER: I beg to differ. They have a great deal to do with ICSI, because we have not addressed one issue of overwhelming importance. In fact, I'm surprised that you hadn't brought it up yourself during your show.

YOUNGBLOOD: And what might that be?

FRANKENTHALER: Whether in the future *fertile* people might also resort to ICSI . . . rather than just those with infertility problems.

YOUNGBLOOD: All right . . . I'll nibble on your bait. Why would normal people—

FRANKENTHALER: I didn't say "normal." I said "fertile."

YOUNGBLOOD: I stand corrected. So why would fertile people spend thousands of dollars to produce a baby without sexual intercourse, when they can do it without cost and much more pleasurably in the usual way?

FRANKENTHALER: Fair enough. But that is why I asked whether you had any children and commented on your age.

YOUNGBLOOD: What's my age got to do with it?

FRANKENTHALER: For argument's sake, let me assume that you have no children, that you are now thirty-five, and that you do wish to have some children in the future . . . but not now. If you wait for another five years—

YOUNGBLOOD: I have now stopped nibbling on your bait.

FRANKENTHALER: Relax! I was trying to generalize from a hypothetical Isabel Youngblood. Given that hypothetical age, just postponing childbearing from thirty-five to forty years of age would at least quadruple your chances of giving birth to a baby with Down's syndrome.

YOUNGBLOOD: I'm fully aware of that. Which is one of the reasons why pregnant women in that age group often have an amniocentesis.

FRANKENTHALER: Of course. But why? Because they are prepared to consider an abortion to terminate such a pregnancy in case an extra chromosome is discovered during such genetic testing. But as we discussed earlier on in your program, with ICSI, several embryos are available—

YOUNGBLOOD: Sure. But for that, the woman would first have to undergo a course of hormone-induced superovulation to generate a supply of eggs. By no means a trivial procedure.

FRANKENTHALER: Yet something that millions of women have done—

YOUNGBLOOD: Because they suffered from infertility problems and IVF was their only option. I thought we're talking about fertile women.

FRANKENTHALER: So we are: about fertile women, who do not necessarily wish to wait for the third month of pregnancy to take the amniocentesis and potential abortion route. Such women could select preimplantation genetic diagnosis: looking for the same genetic markers as in amniocentesis, but doing so with a three-day-old embryo prior to transfer of the embryo into her uterus. Given such a scenario, she would simply discard such a defective embryo and use another one. As I already pointed out earlier in the program, we are then not talking about abortion but selective reduction of the risk. For many women . . . especially those who do not consider a three-day-old embryo in a Petri dish a baby . . . this is not just a semantic quibble . . . because she might feel differently about a three-month-old fetus in her uterus. And as professional women postpone child bearing to that more dangerous age, some of them will consider their access to pre-implantation genetic diagnosis a sufficient incentive to also undergo superovulation.

YOUNGBLOOD: And that will cause millions of older women—

FRANKENTHALER: Just call them mature women.

YOUNGBLOOD: Whatever! To stand in line at ICSI clinics for getting their babies?

FRANKENTHALER: I didn't say millions. I don't know how many, but I believe the number will surprise all of us . . . at least in the affluent countries, notably Europe and perhaps also Japan . . . where the average number of children is below 1.5 per family. It's an option they should not be denied. Especially as the rapid advances in genomics will allow us also to screen for certain other genetic markers . . . say a high predisposition to certain cancers.

YOUNGBLOOD: And where will that stop . . . once we start looking for genetic markers? First one marker . . . then several . . . and then a lot of markers? What kind of slippery slope are we then following?

FRANKENTHALER: In time, it may become quite slippery. Another reason why this topic ought to be discussed and debated. But making it illegal is unlikely to work, other than stimulating medical tourism . . . with its inherent discrimination in favor of the wealthy who can afford to indulge in such luxury.

YOUNGBLOOD: So it's the possibility of pre-implantation genetic diagnosis that will drive fertile women to consider in vitro fertilization?

FRANKENTHALER: That might be the reason now . . . today . . . without further technical advances. But consider the fact that we now know how to store sperm and embryos for years—

YOUNGBLOOD (*laughs*): And "cryopreservation" will become such a standard word that even cooks will talk about "cryopreservation" of food in the kitchen?

FRANKENTHALER: Think about the day when we shall know how to store eggs or ovarian tissue in that manner . . . a scenario we already discussed. It's almost here. I'd give it another decade or two and it will start becoming routine.

YOUNGBLOOD: And?

FRANKENTHALER: And then offer young—not just mature—women the choice to undergo superovulation in their early twenties, when the quality of their eggs is far superior to those in their middle or late thirties.

YOUNGBLOOD: And store their young eggs for eventual use by IVF years later?

FRANKENTHALER: Eggs . . . or ovarian tissue. And do it as a reproductive insurance policy . . . for young women embarking on ambitious careers, who are not prepared to choose that early between motherhood and career.

YOUNGBLOOD: With their eggs in the bank, they might as well then get sterilized at that young age rather than practice contraception . . . and just check out an egg when they're ready for motherhood. Is that it? The end of the Pill and other contraceptives?

FRANKENTHALER: I'm sure that some will do that.

YOUNGBLOOD: And why not men?

FRANKENTHALER: And men, of course. Especially since the technical aspects of cryopreserving sperm for years or even decades has already been solved.

YOUNGBLOOD: And you think that's good?

FRANKENTHALER: It's neither good . . . nor bad. It's an option that should be available to those women . . . and men . . . who wish to exercise it.

YOUNGBLOOD: I don't think we are talking science anymore . . . or ICSI.

FRANKENTHALER: You're right. We're talking about women's choices . . . or religion. That's why it's time for me to leave. As a physician, I can present couples with technical facts, but not with making the most important decision in their lives. (*As he rises.*) But would you mind if I offered one last piece of advice?

YOUNGBLOOD: That depends on what it is.

FRANKENTHALER: Let me propose a theme for one of your future programs.
YOUNGBLOOD: Okay. Let's hear it.
FRANKENTHALER: "Are sex and fertilization ready for divorce?"
YOUNGBLOOD: I gather you think they are.
FRANKENTHALER: My opinion is not important. I'm talking about future generations . . . starting with those who right now are in school. They should debate that issue. Because they are the ones who will indulge in sex in an age of technological reproduction.

(*End.*)

Taboos

A Play in Two Acts

Prelude

My first play, *An Immaculate Misconception* (by now translated into eleven languages), illustrated in the form of "science-in-theatre" some of the ethical dimensions of the use of one of the most exciting assisted reproductive techniques, ICSI, which is also the subject of the preceding pedagogic wordplay in this volume.

Five plays later, in *Taboos,* I return to the topic of sexual conduct in an age of technological reproduction, which can also be described as the complete separation of sex and reproduction. Sex—motivated by love, lust, or curiosity—will no doubt continue as usual, while reproduction will increasingly occur under the microscope or by other "alternative" means. But instead of focusing, as I did in *An Immaculate Misconception* and in *ICSI,* on the technical "yang" of this theme, I now turn in my sixth play to the social "yin" with its much more subtle and complex components. As Chinese cosmology proclaims, only a combination of yin and yang produces all that comes to be, in other words the next generation of persons and of ideas.

Terms such as "marriage," "family," and "parent" used to have firm denotations. They were the rock on which our cultural values rested. Terms such as "embryo," "baby," or "twin" were also considered unambiguous. Assumptions that marriage must be heterosexual and that a child cannot have two parents of the same sex were never even considered assumptions, because they were beyond questioning.

All of these terms have become destabilized, their meanings blurred, their ranges extended. Some would blame in vitro fertilization technology during the past three decades for these developments, but in actual fact major social and cultural changes—primarily in the United States and Europe—were even more responsible for the monumental shift that has caused so much fear and antagonism, especially among the ever increasingly strident fundamentalists in the United States. So why not write a play about a situation where "family" and "parent" have assumed disturbingly fuzzy meanings? This is why I have deliberately situated *Taboos* in two of the most socially and politically polarized parts of the United States: the San Francisco Bay Area and the American Deep South.

Even though I have spent half my life in the San Francisco Bay Area, I do not wish to be regarded as a proselytizer for either of the two extreme views I present in *Taboos*. That is why I end the play on a biblical note, emphasizing the need for compromise in a situation where there can be no winners. I trust that the play's ending, as well as some of the serious speeches about motherhood presented through the mouth of the Christian fundamentalist character, will indicate my sympathy with some of her views, though, admittedly, there is a touch of satire in the representations of all of my characters. I wrote *Taboos* mostly in London, but early parts also in Ireland and Germany, as an American agent provocateur born in Europe who has rediscovered his European roots and acquired a more distanced as well as more nuanced view of America. Unquestionably, agent provocateur is the role that suits me best as a late-blooming playwright, because the issues that interest me most are intrinsically provocative as well as complex. And few topics are as provocative and complex as the present questioning of the social meaning of parenthood and family, where every horror projection can be countered with a "But what if?" scenario. That is why in *Taboos* I have mostly taken the yin side of the argument.

Cast

SALLY (SYDNEY) PARKER	33 years old, anchor woman for a San Francisco television station, later the mother of Tucker.
CAMERON PARKER	35 years old, brother of Sally, conservative, church-attending certified public accountant in Mississippi. He speaks with a pronounced Southern accent.
PRISCILLA PARKER	33 years old, Cameron's wife, highly conservative and religious housewife, later the mother of Ashley. She also speaks with a pronounced Southern accent.
DR. HARRIET CAROTHERS	37 years old, urologist in San Francisco, later the mother of Jan.
MAX CAROTHERS	32 years old, brother of Harriet, lawyer in the public defender's office, San Francisco.

Time

The present.

Location

Mostly in San Francisco, occasionally in Mississippi.

Act 1

Scene 1

(*The dining room of* SALLY *and* HARRIET*'s apartment in San Francisco. A rectangular, preferably expandable, dining room table is in the center* *of the room. It has been set formally and festively, including champagne (still empty) and wine (partly filled) glasses. The centerpiece, a silver tray with a large elongated object wrapped in gold foil, remains undisturbed. Except for a fourth, untouched setting, it is clear that the meal is nearly over.*

HARRIET, festively dressed, is seated at the table with a glass of wine in one hand and her cell phone in the other.)

HARRIET (*into cell phone*): You'll be okay. Just drink plenty of water. (*Pause.*) At least a couple of liters. (*Pause.*) No . . . of course not all at once. (*Pause; then as* SALLY, *smartly dressed, enters from the bedroom:*) I've got to hang up. Just make an appointment during office hours. (*She hangs up and turns to* SALLY.) Sorry about that.

SALLY: You promised. No more patients. Not tonight. Unless you consider me one.

HARRIET: I know, I know. Look, I'm switching it off! Look.

(*She switches the phone off. They embrace and kiss.*)

Sexy baby . . . (*Beat.*) You ready?

SALLY: I guess so.

HARRIET: Sure you still want to go through with it?

SALLY: Of course. Why wouldn't I? It's . . .

HARRIET: Exciting.

SALLY: It's more than that.

HARRIET: Have a drink.

SALLY: No.

HARRIET: You can have one drink.

SALLY: I don't need a drink, I just . . .

HARRIET: It's Cameron, isn't it?

(SALLY *nods.* HARRIET *strokes her hair.*)

You want to wait a bit longer?

SALLY: Oh, what's the use? My brother obviously thinks I'm a lost cause.

HARRIET: Maybe his plane got delayed.

SALLY: Then why didn't he call? I just thought we'd finally be able to . . .

HARRIET: I'm sorry, baby. We're now your family. Me and Max. You can even include my parents.

(SALLY *manages a sad smile.*)

God, you look great!

(SALLY *cheers up. They kiss.* MAX *enters and interrupts the kiss. He is well dressed in slacks and pressed shirt but no tie.*)

MAX: The pastry chef wishes to know if you are ready for the dessert.

(HARRIET *and* SALLY *break off the kiss.*)

It's my only Southern specialty.

HARRIET: How about waiting a couple more minutes?

MAX (*with a tiny note of impatience*): Well . . . why not. But I better turn off the oven.

(MAX *goes to the kitchen. After brief pause he returns to the dining table, grabs a glass of wine, and flops into a seat. There's a little pause while they wait.*)

SALLY: Damn him!

MAX (*as if raising a toast*): Damn him! Damn Cameron!

(*A pause.* HARRIET *and* MAX *drink.*)

SALLY: You know, this year he sent me a birthday card. My first birthday card since I left home! I took it as a sign.

HARRIET: Not everyone finds it easy to forgive and forget.

SALLY: There's nothing to forgive. Besides, he doesn't even know why I wanted him here.

MAX: Okay, look. He dropped you. I'm sorry. But now . . . why don't we just pop the cork, toast each other, and then get started?

SALLY: You two go ahead. No alcohol for me.

MAX: Oh, come on! Just ask the doctor. (*He points to* HARRIET.) Tell Sally a glass of champagne can't hurt her . . . other than improve her mood.

HARRIET: Max. Just go get the champagne. Let's get on to the main event.

(MAX *exits.* HARRIET *puts music on the stereo and offers* SALLY *a dance.* SALLY *accepts and they dance like a courting couple. At the very least they dance to the first few lines of Cole Porter's "Let's Do It (Let's Fall in Love)."*)

Birds do it, bees do it;
Even educated fleas do it—
Let's do it, let's fall in love.
In Spain, the best upper sets do it, . . .

Taboos | 41

(*As* MAX *returns from the kitchen with a champagne bottle, there is a little rap at the door barely heard above the music.* MAX, *still holding the bottle, answers the door.* CAMERON *stands in the doorway. He is dressed in a conservative suit and tie. His first sight is of his sister and* HARRIET *locked in an embrace.* SALLY *sees him, instinctively pulls away from* HARRIET. *An awkward pause.*)

MAX: Ahem!

SALLY: You came . . . !

(*As* SALLY *rushes at* CAMERON *and hugs him,* HARRIET *goes over to the stereo and stops the music.*)

CAMERON: The doorman let me in. I wanted it to be a surprise.

SALLY: Come in, come in!

CAMERON: I'm sorry I'm so tardy. My plane was late.

SALLY: It's okay, never mind.

(CAMERON *steps into the room,* SALLY *leading him by the hand.* HARRIET *and* MAX *hang back.*)

HARRIET: You want to be alone?

SALLY: No, no. Cameron . . . this is Harriet.

CAMERON (*flustered*): Pleased to meet you, Ma'am.

HARRIET: I'm glad you made it. I've always wanted to meet some member of Sally's family. She's talked about you a lot lately.

(HARRIET *and* CAMERON *shake hands, with evident coolness.*)

Too bad you missed the first two courses. We'd finally given up on you.

MAX: So you're the mysterious brother.

SALLY: This is Max.

CAMERON: Pleased to meet you, Sir.

HARRIET: Pop the cork, Sir Max.

(MAX *takes the champagne to the table.*)

SALLY: Not me. Later . . . maybe.

HARRIET (*she offers a full glass to* CAMERON): You must join us in a toast to Sally and to this special occasion.

CAMERON (*embarrassed*): Much obliged, Ma'am, but . . .

HARRIET (*interrupts*): You can call me Harriet.

CAMERON (*even more embarrassed*): I can't . . .

HARRIET: Of course you can.

CAMERON (*points to glass*): I mean . . . I don't indulge.

(MAX *and* HARRIET *exchange surprised glances.*)

SALLY: Cam doesn't drink.

MAX: Well, can I get you something else? Orange juice? Sparkling water? Ginger ale?

CAMERON: Any Coke?

MAX: I didn't get Coke. (*Grins at* SALLY.) Too much phosphoric acid, according to Sally. You know your sister, obsessed with everything organic . . .

HARRIET (*joins in*): Hates cell phones . . .

MAX: Never a hair out of place . . .

SALLY (*laughing*): Enough compliments! Cam and I will both have ginger ale. Color coordination!

(MAX *hurries into the kitchen.* SALLY *hugs* CAMERON *again.*)

I'm so pleased you came!

CAMERON: Me too. It's been a long time. But Jeez . . . You sound so different . . . I mean I've seen one of your broadcasts on the Internet. Folks back home would never guess you're one of us. (*Beat.*) But what's the special occasion?

HARRIET: Sally and I are . . .

SALLY (*interrupting*): First things first. You must be hungry.

CAMERON: That's okay. I'll join you for dessert.

(HARRIET *suppresses a giggle.* SALLY *glares at her.*)

SALLY: Well, sit down.

(CAMERON *takes a seat.*)

CAMERON (*addressing* HARRIET): So Ma'am, what do you do? (*Beat.*) I mean, Harriet.

HARRIET: I'm an urologist (MAX *enters with two ginger ales.*) and you'd be surprised how many of my patients call me "Ma'am."

MAX: Especially men with their pants around their ankles.

SALLY: Max, cut it out!

HARRIET (*with glass raised to* CAMERON): May this be the first of many visits! (*To* SALLY *and* MAX.) Success at the first try!

SALLY (*with glass raised to* CAMERON): I'm awfully glad you're here. (*Then raising glass to* HARRIET *and* MAX.) Success!

MAX: Success!

(*All three turn to* CAMERON, *waiting for his toast, but he's tongue-tied.*)

CAMERON: Jeez . . . I don't know what to say.

MAX: Anything that comes to your mind.

CAMERON: I have no idea what y'all are celebrating. . . . Sally just wrote it was a surprise. But . . . (*He raises his glass to no one in particular.*) . . . whatever it is, may the good Lord bless it . . . and you.

SALLY (*she leans over and gives* CAMERON *a kiss on his cheek*): Nice toast, Cam. Cheers!

(*They all join in as they clink glasses all around, being careful that every glass is being touched, and sip briefly from their glasses, the conversation ceasing somewhat awkwardly.*)

I don't think I can stand waiting much longer.

HARRIET: Well then, let's proceed. (*She reaches into her pocket to produce what appears to be a gold coin and tosses it to* MAX.) Catch!

(MAX, *unprepared for the toss, fails to catch the object, which falls on the table or floor near* CAMERON, *who picks it up.*)

CAMERON (*surprised at its lightness*): Chocolate?

(CAMERON *hands it to* MAX.)

MAX: Hardly.

(MAX *starts to move toward the bedroom door, a little nervously.*)

SALLY: Oh, and here's this DVD. Remember . . . ?

MAX: Aye, aye, captain.

(MAX *takes the DVD and exits.*)

HARRIET (*barely able to suppress her amusement*): Sally . . . think of something suitable to say while we wait.

CAMERON: Wait for what?

HARRIET: Sally . . . please put the poor man out of his misery.

SALLY: Why don't you go check about the dessert?

HARRIET: Yes, darling.

(HARRIET, *with a droll smile, goes into the kitchen.* CAMERON *watches her go. There is a momentary pause.*)

CAMERON: Was that a condom?

SALLY: Of a sort.

CAMERON: What does that mean?

SALLY: Never mind. (*Beat.*) You don't like Harriet, do you?
  (CAMERON *shrugs.*)
  You disapprove of us.
CAMERON: You know I do.
SALLY: Why?
CAMERON: You know why. It's unnatural.
SALLY: You sound exactly like Mom and Dad . . . for whom any word starting with "homo" is an abomination . . . with the possible exception of "homo sapiens" . . . and that only if you make it plain it has nothing to do with evolution.
CAMERON: They only did what their conscience made them do.
SALLY: Like not inviting me to their son's wedding?
CAMERON: You know that wasn't my fault. Mom said she wouldn't come if you . . .
SALLY: And of course we never disobey our parents, do we?
CAMERON: Come on, Sid, I'm trying.
SALLY: Don't call me Sid. Those days are gone.
CAMERON: Okay . . . Sally. (*Beat.*) Hey, you remember the time we went to the costume party at the McNultys' house?
SALLY: It was a blast!
CAMERON: And you went dressed as Fidel Castro . . .
SALLY: False beard and exploding cigar.
CAMERON: It was funny . . . but also weird.
SALLY: Not everybody thought it was funny. I was grounded for a week for that stunt.
CAMERON: You were always interested in stuff like that.
SALLY: Stuff like what?
CAMERON: Y'know . . . stuff. Being different.
SALLY: Different? Maybe back home. But this is San Francisco, Cam, not Mississippi! (*Beat.*) Gays have rights here.
CAMERON: Maybe here.
SALLY: Here? Listen . . . gay partnerships or marriages are becoming legal in lots of places. Belgium, the Netherlands, Canada, South Africa . . . even England. (*Beat.*) Or take Catholic Spain where they've legalized adoption by gay couples.

CAMERON: I bet the Pope doesn't approve.

SALLY: Oh, Cam! Just exactly what's wrong with two loving women adopting a child . . . or even better, giving birth to one?

CAMERON: Hold on a minute. What are you saying?

SALLY: I'm saying, what's wrong with a gay woman giving birth?

CAMERON: Is *that* what you're aiming to do?

> (CAMERON *waves his hand around the room.* SALLY *is silent.*)
> Sally! Is it?
> (*At this moment* HARRIET *comes out of the kitchen.*)

HARRIET (*knocks on the bedroom door*): Is the coast clear?

MAX (*offstage*): Come in.

> (HARRIET *goes into the bedroom.* CAMERON *stares at* SALLY, *then turns to look at the object on the table.*)

CAMERON: Jeez. Is that that what this ceremony is all about?

SALLY: Cam. Let me ask you something. What do you think about artificial insemination?

CAMERON: I don't believe in artificial procreation.

SALLY: Not even for a married couple?

CAMERON: You're not a married couple.

SALLY: That wasn't my question.

CAMERON: Well . . . okay. I can see how, for a married couple, there might be . . . circumstances that would excuse it, in God's eyes. I guess.

SALLY: So you're not against it, in principle?

CAMERON: I guess not . . . under special circumstances.

SALLY: And you don't think my relationship with Harriet is just such a circumstance?

CAMERON: No.

SALLY: Why not?

CAMERON: You aren't infertile, are you? If you want a baby, settle down with a decent man and . . .

SALLY: I don't want a man. I never have and I never will. (*Beat.*) Do you believe in love?

CAMERON: You know I do.

SALLY: I love Harriet. I want a baby. With Harriet. Not with a man.

CAMERON: Jeez, Sally! I don't know . . . there are just . . . some things . . .

SALLY: What things? In Mississippi, maybe. That's why I moved here. (*Beat.*) So I could keep on loving you *and* love Harriet too.

CAMERON: I don't know how to say it . . . but in God's eyes it can't be right.
(*Enter* HARRIET *from the bedroom, followed by* MAX. HARRIET *starts to unwrap the gold foil on the dining table, disclosing a large syringe.* CAMERON *is surprised.*)
A turkey baster?

HARRIET: We've got a fully equipped kitchen.

CAMERON (*to* HARRIET): Y'all trying to shock me?

HARRIET: No. But it wouldn't be difficult. Actually, this one is too small for a Thanksgiving turkey. (*To* MAX.) I'll take it. (*She points to handkerchief in his hand containing the condom. To* SALLY.) Ready?

SALLY (*to* CAMERON): I'll be back in a jiffy. (*To* MAX.) Be nice to him . . . try some guys' talk.
(HARRIET *heads for the bedroom with the "turkey baster."*)

HARRIET: You two might as well get to know each other while we're busy. It won't take long.
(*As* HARRIET *leaves, with her back to the men so they can't see what she's doing, she carefully opens the handkerchief and removes the condom, which she holds at the top.*)

CAMERON (*pointing first to the center of the table and then to the door through which* SALLY *and* HARRIET *had exited*): You mean they're going to use that thing?

MAX: I guess Sally didn't tell you what the occasion was.

CAMERON: In her first letter in years, she begged me to come. She thought I'd bring her luck on the most important day of her life. (*Beat.*) I thought she was getting married . . . or engaged . . . or something.

MAX: That's exactly what it is . . . quite some . . . thing. Sally quit as our top local TV news anchor for this! And that little chocolate you saw in my hand? Yes, that was a condom. But a special one, without spermicide, which we wrapped ourselves. Actually, we've got all kinds here in the Bay Area, we use them a lot. The whole range: pagoda-shaped . . . ribbed . . . flavored . . . every color under the rainbow. By now they've filled what you called the turkey baster with my seed and are probably all finished. It's faster than intercourse, no foreplay . . .

CAMERON: Thank you! That's enough information.

MAX: Just trying to help.

CAMERON: And I suppose you also consider that . . . natural?

MAX: As far as fertilization is concerned? (*He shrugs his shoulders.*) It's just a question of delivery vehicle.

CAMERON: That's all you see yourself as? A delivery vehicle?

MAX (*chuckling*): Yeah. Like UPS. Or FedEx in my case. I'm known to be fast if I concentrate . . .

CAMERON: Listen, bud. I realize y'all think I'm a Southern hick, but why don't you just cut it all out.

MAX: My apologies! No brother of Sally's could be a hick. I shouldn't have made that crack.

CAMERON: Well, okay, then.

(*An awkward pause.*)

So . . . er . . . Max. You married?

MAX (*shaking his head*): Maybe later.

CAMERON: How old are you?

MAX: Thirty-two.

CAMERON: Why postpone it?

MAX: I've still got a humongous student loan to pay off. First college . . . then law school.

CAMERON: What sort of lawyer are you?

MAX: I'm in the Public Defender's Office. Indigent clients. So you can imagine how long it will take me to pay off my loans. (*Beat.*) So you're a CPA.

CAMERON: That's right.

MAX: A number cruncher.

CAMERON: I like to keep people honest . . . especially with tax returns.

MAX (*genially*): Must be good, steady business.

CAMERON: I can't complain.

MAX: How long have you been married?

CAMERON: Over four years.

MAX: Any kids?

CAMERON: Not yet, but still trying.

MAX: That's pretty tough. You thought about IVF?

CAMERON: No.
MAX: You know, you really should.
CAMERON: I put my trust in the Lord.
MAX: Well, it's your decision.
CAMERON: No, sir! God will decide.
MAX: Okay. Good luck. You'll need it.
CAMERON: Thank you.

(*There is a long, awkward silence.*)

MAX: So you're from Europa, Mississippi.
CAMERON: Eupora . . . not Europa.
MAX: An anagram of Europa?
CAMERON: You think we use anagrams to name our towns? (*Beat.*) Have you ever read any Syriac literature?
MAX (*straight-faced*): I do every morning . . . right after my Yoga.
CAMERON (*calling his bluff*): You mind sharing what you read today?
MAX: You got me. So what's Syriac?
CAMERON: Ancient Aramaic . . . and still the liturgic language of some Eastern Christian Churches.
MAX: Don't tell me you read Syriac.
CAMERON: I researched it, because I wanted to know where the name of my hometown came from. In Syriac, Eupora was the old name of Corinth. (*Beat.*) Corinth in Greece . . . not our Corinth in Mississippi.
MAX: Okay, but why so many Greek names in the south? Athens, Georgia, Troy, Alabama . . .
CAMERON: I reckon its all about slavery. The Greeks condoned it and didn't let their slaves vote, and so did our forefathers. Also the reason for all the Greek columns in our plantation homes.
MAX: That makes sense. So it's Eupora.
CAMERON: Sally and me were raised there. Nice town . . . but too small for a CPA. That's why my wife and I moved to Jackson. (*Beat.*) So . . . how did you get to know Sally?
MAX: Through Harriet.
CAMERON: I see. (*Pause.*) And how long have you known Harriet?
MAX: All my life.
CAMERON: No kidding?

MAX: Yeah.

(CAMERON *puzzles over this for a moment. Then, walking right into it:*)

CAMERON: How come?

MAX: I'm her brother.

(CAMERON *is stunned.*)

CAMERON: What did you just say? (*Beat.*) You're Harriet's brother?

MAX: Yeah.

CAMERON: I've had it.

(CAMERON *stands up.*)

MAX: You're leaving?

CAMERON: You bet I am.

MAX: But you just got here.

CAMERON: Nevertheless, I've gotta go.

MAX: Stay for dessert at least. I spent over an hour on it.

CAMERON: I can't help that.

MAX: Why are you leaving?

CAMERON (*turning on him*): You see, Max, to me . . . what you're describing is incest.

MAX: Incest? Listen! I'm Harriet's brother . . . not Sally's.

CAMERON: You've just had sex with your sister-in-law!

MAX: I think you're confusing genetics with some hang-up of yours.

(CAMERON *goes to leave.*)

Hold it bud! At least try some dessert! Pecan pie . . . my only Southern specialty. I did it for Sally's sake.

(CAMERON *hesitates by the door.*)

You come all this way, and now you're going to leave, without saying goodbye to your sister?

(CAMERON *is sufficiently challenged by this to be halted in his tracks. He listens.*)

You know, I bet they'll make it on their first try. Before you know it, you'll be an uncle.

CAMERON: God help me.

MAX: They're the best organized couple you ever saw. For the last three months, they've determined Sally's optimum fertile time practically to the hour. With a fancy fertility monitor—computer-driven—and a drop of urine one can now pinpoint the most fertile day of the month. Just a question of measuring the relevant hormone levels.

CAMERON: Jeez!

MAX: So you see, they're not just a couple of dykes acting out some dumb fantasy. They know what they're doing. And especially my sister; always the hi-tech doctor.

CAMERON: I don't wanna hear any more!

MAX: Do you believe in the idea of the family?

CAMERON: I sure do.

MAX: You believe in keeping things in the family?

CAMERON: Yes, I do.

MAX: Well, that's what they're doing in there. You think it's better to have a completely anonymous sperm donor, someone you never met, someone you know nothing about?

(CAMERON *is stumped for an answer.*)

CAMERON: I don't know what to think. My mind is doing somersaults here.

MAX: Well, that's a start. Now keep it up.

(*A pause.* HARRIET *and* SALLY *emerge from the bedroom.*)

HARRIET: Mission accomplished!

(*End of scene 1.*)

Scene 2

(*Three weeks after scene 1.* PRISCILLA *and* CAMERON's *living room in Jackson, Mississippi.* PRISCILLA *enters, wearing casual clothes for domestic work and perhaps rubber gloves. She's carrying a large old cardboard box. She puts it down on the floor and kneels down to open it. She rummages around inside and finds something. It's a little girl's doll—a Barbie doll or a rag doll, somewhat old and battered. She studies it, fussing over it for a moment.*)

PRISCILLA: O Lord, give me a sign. Give me a sign, O Lord.

(PRISCILLA *stops as she hears* CAMERON.)

CAMERON (*offstage*): Honey?

(PRISCILLA *puts the doll away and closes the lid of the box.*)

PRISCILLA: In here.

(PRISCILLA *stands up and crosses to the table as* CAMERON *enters. He's carrying two Bibles.*)

CAMERON: What's in the box?

PRISCILLA (*ignoring that question*): Ready?

(*They sit down together, taking a Bible each. There is a brief pause as* PRISCILLA *takes her gloves off.*)

CAMERON: Well?

PRISCILLA: Yes?

CAMERON: It's your turn to read.

PRISCILLA: Never mind. I want to hear you do it first.

(*There's just a hint of tension in* PRISCILLA*'s remark, which* CAMERON *notices. He opens the Bible and starts browsing through, looking for a lesson. She takes the book from him, finds a page, and indicates for him to read.*)

CAMERON: "When Isaac was forty years old, he married Rebekah, the daughter of Bethuel the Aramean and the sister of Laban. Isaac pleaded with the Lord to give Rebekah a child because she was childless. So the Lord answered Isaac's prayer, and his wife became pregnant with twins."

(PRISCILLA *takes* CAMERON*'s hand in hers and they pray.*)

PRISCILLA: Answer our prayers, O Lord.

CAMERON: Lord, answer our prayers. Make my wife Priscilla pregnant.

PRISCILLA: I'm begging you, O Lord, to give me a sign that I may become pregnant. Make me pregnant, O Lord. Give me a baby, I'm begging you.

CAMERON: And yet Thy will be done, O Lord. Amen.

(PRISCILLA *stares at* CAMERON *as if he had just hushed her.*)

PRISCILLA: Okay. And now it's Forgiveness Time. You first. What would you like me to forgive?

CAMERON: Well, um . . . let me think.

PRISCILLA: You could start by explaining why you lied to me.

CAMERON: I didn't, Prissy.

PRISCILLA: Yes, you did.

CAMERON: I've never lied to you.

PRISCILLA: Why didn't you tell me you'd gone to see her? Flying all the way across the country . . .

CAMERON: Because you would've argued with me.

PRISCILLA: No . . . I would've told you not to see that sinner.

CAMERON: She's my sister.

PRISCILLA: In the Lord's eyes, she's a sinner. Your parents would've said the same.

CAMERON: Don't bring them into this.

PRISCILLA: Why not? You should have more respect for your Mom and Dad.

CAMERON: I do. (*Beat.*) All right. I ask your forgiveness for . . . for not telling you I went to San Francisco.

PRISCILLA: Why did you go?

CAMERON: She asked me to.

PRISCILLA: Did you ask the Lord what you should do?

CAMERON: I always do . . . you know that.

PRISCILLA: I don't believe you.

CAMERON: Do I get your forgiveness or not?

PRISCILLA: Not yet. I need to hear you say that woman is a sinner!

CAMERON: All right. She's a sinner. But . . .

PRISCILLA: But what? No buts!

CAMERON: "Lord, make us ever mindful of the wants and needs of others."

PRISCILLA: Are you praying for her wants and needs? If you mean what I think you mean, that's creepy. (*Beat.*) Why did you go see her now?

CAMERON: I wish I'd done it earlier.

PRISCILLA: You mean she's repenting?

CAMERON: It's not for me to say.

PRISCILLA: She's living in sin and will never enter the kingdom of heaven!

CAMERON: How do you know she's living in sin?

PRISCILLA: She's living with another woman!

CAMERON: Lots of people are living together . . . men with women, women with women, men with men . . . without sinning.

PRISCILLA: Unmarried and not sinning?

CAMERON: They consider themselves "domestic partners."

PRISCILLA: What's that supposed to mean?

Taboos | 53

CAMERON: The legal definition says: "Domestic Partners are two adults who have chosen to share one another's lives in an intimate and committed relationship of mutual caring." That's a quote.

PRISCILLA: Where did you get that from?

CAMERON: Max.

PRISCILLA: Who's Max?

CAMERON: Harriet's brother. He's a lawyer.

PRISCILLA: So you've become buddies . . . Max and you? And Sally and that woman aren't sleeping together?

CAMERON: I don't know . . . I didn't ask her.

PRISCILLA: You see?

CAMERON: See what? They live in a nice place.

PRISCILLA: Oh, really? Did you ask her what they do in their bedroom as part of a . . . (*She assumes a nasty, sarcastic tone.*) "committed relationship of mutual caring"?

CAMERON: How can I ask that? What would you say if someone asked what we do in our bedroom?

PRISCILLA: It's none of their business. We're married!

CAMERON: Married people do sinful things in bed.

PRISCILLA: Cameron!

CAMERON: Orthodox Jews are only permitted sex in the missionary position. You, Pris, always want to be on top. Not just lie on top and move slowly. *Sit* on top . . . and bounce. And then you complain that I come too soon.

PRISCILLA: Cameron Parker! Hush! We aren't Jews.

CAMERON: Okay, okay. What about when I asked you to try . . .

(CAMERON *stops.*)

PRISCILLA: Try what?

CAMERON: You remember.

PRISCILLA: I remember nothing.

CAMERON: Come on . . . you know. (*Beat.*) Use me like a lollypop.

PRISCILLA: That's icky! And you aren't exactly vanilla-flavored! Cameron! What's come over you? You're different since you came back from San Francisco.

CAMERON: I haven't changed. I'm just starting to see things differently.

Sally . . . and Harriet are nice women, who care for each other. I don't know what else they're doing . . . but it isn't Christian to deny your only sister.

PRISCILLA: Tell that to your parents. See how they like it.

CAMERON: Maybe I will. (PRISCILLA *glares at him*.) But not now. (*Beat*.) Now it's your turn.

PRISCILLA: I haven't forgiven you yet.

CAMERON: Aw, Jeez, Prissy!

(CAMERON *gets up to leave*.)

PRISCILLA: Where are you going? Don't you leave this table until we're finished!

CAMERON: If you're not going to forgive me, then Forgiveness Time is over!

PRISCILLA: All right. Forgiven. My turn now.

(CAMERON *sits back down*.)

CAMERON: Go ahead. I'm ready to forgive you.

PRISCILLA: Thank you. (*Beat*.) While you were away . . . I had . . .

CAMERON (*a slight feeling of alarm*): What?

PRISCILLA: I had . . . thoughts.

CAMERON (*relieved*): Thoughts?

PRISCILLA: Yes.

CAMERON: Thoughts about what?

PRISCILLA: Adultery. (*Beat*.) I had thoughts about committing adultery.

CAMERON (*astonished*): My gosh! Anyone in particular?

PRISCILLA: No, of course not.

CAMERON: So you didn't do anything about these thoughts?

PRISCILLA: Cameron!

CAMERON: I'm just asking, so I can get the facts.

PRISCILLA: The facts are we don't have a baby!

CAMERON: Well . . . I'm doing everything I can.

PRISCILLA: I know, honey. (*Beat*.) Please forgive me.

CAMERON: I forgive you.

PRISCILLA: I'm much obliged.

(PRISCILLA *kisses his cheek*.)

CAMERON: Anyway . . . if you're considering adultery . . . you must be blaming me. How do you know it's my fault?

PRISCILLA: Because it must be.

CAMERON: Because I'm the sinner in the family? Because I want to do it from behind?

PRISCILLA: Cameron. We aren't dogs!

CAMERON: And that makes me a sinner?

PRISCILLA: I . . . I don't know! I just want an answer! I need an answer.

CAMERON: All right. I'm sorry. How about I go see a doctor?

PRISCILLA: I thought you didn't want to.

CAMERON: Maybe the Lord now wants me to go see a doctor.

PRISCILLA: Okay. When?

CAMERON: As soon as you do the same.

PRISCILLA: I don't want to go.

CAMERON: You want a baby?

PRISCILLA (*she closes her eyes*): O Lord, please tell me what to do. Should I go see a doctor or should I wait for a sign?

(PRISCILLA *listens for a second in prayer, then opens her eyes.*)

CAMERON: Well?

PRISCILLA: I'll go see a doctor.

CAMERON: Okay, then. Anything else?

PRISCILLA: No, Forgiveness Time is over.

CAMERON (*standing up*): Hey, isn't that box from the attic? What are you doing with that?

(CAMERON *goes over to the box and starts opening it.*)

PRISCILLA: It's just a few old things.

CAMERON: You planning a yard sale? We don't need the money . . . (*He peers inside.*) Jeez, will you look at that. Your old dolls.

PRISCILLA: Yes.

CAMERON: What are you doing with these?

PRISCILLA: Nothing. I just thought . . . maybe . . .

(*The phone rings.*)

CAMERON: Oh, let me get that, it's probably Jimmy about golf on Saturday.

(CAMERON *goes to the phone.* PRISCILLA *goes over to the box and takes out another doll. Unseen by him, she holds it close to her chest and listens as he talks into phone.*)

Parker residence . . . Oh, hi . . . hi, yes, I'm . . . I'm fine . . . we're fine . . . yes, she's right here . . . No, that's okay, we can talk . . .

(*Now* PRISCILLA *realizes who is on the phone. She puts the doll down, tense.*)

Um . . . yes. Yes, I can . . . Yes . . . Why? . . . Oh! Oh, Jeez . . . Oh, my gosh . . . on the first try? That's awesome. I see . . . congratulations. Okay. Well. That's . . . that's . . .

(CAMERON *becomes aware of* PRISCILLA'S *presence.*)

That's . . . that would be no problem for me . . . No, no problem. Leave it to me. When would that be? . . . I see . . . Okay. Well. We'll talk again before that . . . Okay, take care.

(CAMERON *hangs up.*)

PRISCILLA: That wasn't Jimmy.

CAMERON: No.

PRISCILLA: That was her.

CAMERON: Yes.

PRISCILLA: Are you going to see her again?

CAMERON: Yes.

PRISCILLA: Because I have forgiven you? That was for the last visit. You can't forgive ahead of time!

CAMERON: Prissy . . . I've got to see her.

PRISCILLA: When?

CAMERON: In September.

PRISCILLA: Why then?

CAMERON: It'll have been nine months.

PRISCILLA (*information is starting to sink in*): You mean *she's* having a baby? (*A pause.* PRISCILLA *prays.*) Oh, Lord! Here's your faithful servant . . . wanting a child more than anything else . . . and *she* ends up pregnant! (*Beat.*) How?

CAMERON: What do mean "how"?

PRISCILLA: Who's the father? An anonymous sperm donor?

CAMERON: No. It's Max.

PRISCILLA: Who's Max? (*Now she remembers.* PRISCILLA'S *jaw drops.*) Oh, good grief! Max! Good . . . *grief!*

CAMERON: Yes, yes, I know.

PRISCILLA: Adultery!

CAMERON: No, wait a minute! Max isn't married.

PRISCILLA (*outraged*): Well, then . . . fornication with her lesbian lover's brother! The Lord cannot . . . will not . . . permit that to go unpunished. O Lord, do not permit that wickedness to occur without consequences!

CAMERON (*trying to calm her down*): Prissy! Sally didn't fornicate with him. That's not how she got pregnant.

PRISCILLA: How do *you* know that?

CAMERON: It was artificial insemination.

PRISCILLA: That's what she told you!

CAMERON: I was there when it happened.

(PRISCILLA's *jaw drops again.*)

PRISCILLA: And you saw *that*?

CAMERON: Well, I didn't see everything . . . but I saw the turkey baster and the condom . . .

PRISCILLA: That's enough! That . . . is . . . enough!

CAMERON: Okay, okay, I agree with you. It is . . . wrong. Yes. Okay. Don't get hysterical.

(*Pause.*)

PRISCILLA: Well, you're not going there again. Not without me. You need protection. Your first visit already turned your mind in a sinful direction.

CAMERON: I told you, I agree with you. It's wrong. It's unnatural. It's sinful. That part . . . at least.

PRISCILLA: Every part of it, Cameron! (*She picks up the doll. There is a long silence. He shuffles over to her.*) This one was always my favorite. I got her when I was four. (*Beat.*) When you said, "That part . . . at least" . . . what was that supposed to mean?

CAMERON: I don't know. I just meant . . . you were looking for a sign. Well? The phone rang . . . Sally's pregnant . . . she used artificial means.

PRISCILLA: You think that's a sign?

CAMERON: Maybe. I mean, if it worked for them . . . maybe it's God's way of showing us He wants us to consider artificial means. It's possible, isn't it? You've got to admit it's possible.

(*Pause.*)
PRISCILLA: If you're going to see her, I'm going with you.
CAMERON: Well, if you're sure you want to . . . but it may not be what you're expecting.
(PRISCILLA *cuddles the doll. Lights out.*)
(*End of scene 2.*)

Scene 3

(SALLY *and* HARRIET's *apartment, ten months later.* SALLY's *baby is just a few weeks old. The stage is empty as the lights go up. The doorbell rings.* SALLY *comes out of the bedroom and answers the buzzer.*)
SALLY: Come on up.
(SALLY *adjusts her clothing and her hair. She takes a look around the room and quickly puts a couple of magazines out of sight. There is a knock at the door. She opens it to* CAMERON *and, a little way behind,* PRISCILLA.)
Come in, come in.
(SALLY *ushers them in. As if making a show of sisterly affection for* PRISCILLA *to see,* SALLY *embraces* CAMERON.)
I'm so glad you could make it. (*To* PRISCILLA.) Priscilla, nice to meet you at last. I'm Sally.
(SALLY *offers a handshake to* PRISCILLA, *which is accepted politely.*)
PRISCILLA: How do you do?
SALLY: Please come in. And let me take your coats.
(CAMERON *and* PRISCILLA *remove their coats,* CAMERON *gallantly helps* PRISCILLA *with hers.* SALLY *collects them and hangs them up while they talk.*)
So did you have a good flight?
CAMERON: You bet.
SALLY: Make yourselves comfortable. Sit anywhere you like.
PRISCILLA: Thank you.
(*As if not wanting to be too comfortable,* PRISCILLA *opts for a dining table chair.* CAMERON *takes the sofa.* SALLY *hovers.*)
SALLY: Would you like a drink, something to eat?

CAMERON: Just some coffee, black.

SALLY: Coffee for you. And what can I get you?

PRISCILLA: Do you have herbal tea?

SALLY: Oh, sure. What kind?

PRISCILLA: Rose hip. We ate on the plane.

SALLY: No problem. Back in a minute.

    (SALLY *exits to the kitchen.* PRISCILLA *takes a good look around as if searching for something.*)

PRISCILLA: Their place seems nice.

CAMERON: What are you looking for?

PRISCILLA: Nothing. I'm just looking.

CAMERON: It's not a house of wickedness.

PRISCILLA: We'll see.

    (PRISCILLA *gets up and walks around the apartment, studying the place like a detective.*)

CAMERON: For heaven's sake, sit down!

    (PRISCILLA *ignores him, carrying on.*)

    You're not going to find anything kinky.

PRISCILLA: No? Then what's this?

    (*She holds up a coffee table photography book. Just at that second, unseen by* PRISCILLA, HARRIET *enters from outside.*)

    Photographs of naked females.

HARRIET: They're not naked, they're nude.

    (PRISCILLA *spins around and sees* HARRIET *in the doorway.* HARRIET *removes her coat. She's wearing a trouser suit, with her hair up. Evidently she has been at work, having brought some paperwork home with her in a briefcase.*)

PRISCILLA: You must be Harriet.

HARRIET: And you must be Priscilla. How do you do?

    (HARRIET *and* PRISCILLA *shake hands, but it's frosty.* CAMERON *stands up and greets* HARRIET *more warmly.*)

CAMERON: Good to see you again.

HARRIET: How was the flight?

CAMERON: Great.

PRISCILLA (*pointing to the book of photographs*): So what is the difference exactly?

HARRIET: Nakedness is something I encounter every day at work. And most of the time it's not an aesthetically pleasing sight. Unlike the pictures in that book, which are nudes, meaning they're beautiful. It's a catalog of an exhibition by Helmut Newton at our Museum of Modern Art. It's still on. Cam could take you there.

PRISCILLA: I wouldn't want to see them.

HARRIET: Suit yourself. All art is a matter of taste. But now that you finally made it here, I hope you'll let go of some of your misconceptions about Sally and me.

PRISCILLA: I don't know about that. But you have good taste in furniture.

HARRIET (*rising to the bait*): Well, that's a start. You know what?

(HARRIET *is about to continue when* SALLY *enters from the kitchen.*)

SALLY: Oh, you're back, that's great. (*To* CAMERON.) The coffee will take another minute.

HARRIET: Coffee? Fantastic! (HARRIET *throws herself down on the sofa.*) I had a bitch of a day.

(SALLY *notices* PRISCILLA *wince at the expression.*)

SALLY: Would you like to see the baby?

CAMERON AND PRISCILLA: Oh, sure!

SALLY: He's in there, asleep.

PRISCILLA: Oh, don't disturb him on our account.

SALLY: No, it's fine. I'll bring him out.

(SALLY *heads for the bedroom.*)

PRISCILLA: You sure?

HARRIET: Don't worry about it.

(*There is an awkward moment as they wait for* SALLY *to return.* HARRIET *rubs her tired face.*)

CAMERON: How's work?

HARRIET: Good . . . but awfully busy.

CAMERON: You know, I'd like to talk to you some time about . . . your work.

HARRIET (*a touch surprised*): You would? Well . . . why not?

(SALLY *returns from the bedroom with a baby wrapped up in a blanket.*)

SALLY (*quietly*): Here. Take a look.

(CAMERON *and* PRISCILLA *step gingerly over to take a peep at the baby.*)

Taboos | 61

CAMERON: Oh, my gosh . . . look at him.
PRISCILLA: He's . . . beautiful.
SALLY: Thank you.
PRISCILLA: He looks like you.
HARRIET: No, he looks like Max.

   (*Stung by this,* PRISCILLA *glares at* HARRIET. SALLY *looks anxiously at* CAMERON.)

SALLY: You want to hold him?
PRISCILLA: Oh, no, thank you.
SALLY: Sure?
PRISCILLA: It's fine. He's yours, you hold him.
SALLY: Cam? You are his uncle.

   (SALLY *offers the baby to* CAMERON. CAMERON *takes him a little awkwardly.*)

CAMERON: Well, hi there, little Tucker. (*Irritated,* HARRIET *gets up and goes to the kitchen. To* PRISCILLA.) Remember, Tucker was our mother's maiden name.
PRISCILLA: Yes, you reminded me.
CAMERON: I did?
PRISCILLA: On the plane. A thousand times.

   (CAMERON *and* SALLY *laugh.*)

CAMERON: It's a neat gesture, calling him Tucker.
SALLY: It wasn't meant as a gesture.
CAMERON: No . . . sure, but . . . he is so great. Look at him, Pris.
PRISCILLA: How are you feeding him?
SALLY: I'm still nursing him.
PRISCILLA: And that's okay?
SALLY: Yes, why wouldn't it be?
PRISCILLA: I meant, with it being . . . not a natural conception.
SALLY: What's that got to do with it?
PRISCILLA: Nothing . . . I guess.
SALLY: He's a normal, healthy, bouncing baby.
PRISCILLA: So he seems.
CAMERON: Yes, you are! Oh, my golly, yes you are!

   (HARRIET *returns from the kitchen with a tray of hot drinks.*)

HARRIET: Refreshments.
   (*The group around the baby splits as* HARRIET *lays out the tray on the dining table.* SALLY *takes* TUCKER *back from* CAMERON *and places him into his crib.* PRISCILLA *edges toward the table to collect her tea.*)
PRISCILLA: I believe that's mine.
SALLY: Sit down, make yourselves at home, for goodness' sake.
   (HARRIET *gets her coffee and sits on the sofa.* SALLY *joins her.* CAMERON *joins* PRISCILLA *on the dining chairs.*)
   Priscilla, why don't you take the armchair?
PRISCILLA: We're fine here.
SALLY: Or sit here?
HARRIET: Sally, don't fuss.
PRISCILLA (*to* HARRIET): So you're the boss in the house, I see.
HARRIET: You mean, am I a butch dyke?
PRISCILLA (*totally shocked*): God forbid I would ever say something like that.
HARRIET: I was just kidding. But no, that's not my role, boss or father. I am Sally's partner. You know, what we're doing here . . . I mean, leaving aside our sexual preferences . . . we're creating a nuclear family. Because that's what Sally craved after losing her own family . . . and what I wanted even though I never lost my own parents and brother! That's what brought us together.
PRISCILLA: And where is the father?
SALLY: He's around.
HARRIET: He comes whenever we invite him. He's mostly an uncle . . .
PRISCILLA: Then he's not much of a father.
CAMERON: Prissy!
SALLY: He's also a very good friend.
HARRIET (*to* PRISCILLA): Excuse me, but what exactly do you mean by "father"? Other than the source of a single, admittedly indispensable, sperm?
SALLY: Harriet, don't.
PRISCILLA: It's okay. I'm very interested to hear her opinions. But you know, to me a father is a whole lot more than just a sperm donor.
HARRIET: Oh, I agree with you. The category "father" embraces all kinds, from loving and involved parents to negligent and abusive assholes.

(*The word hangs in the air.*)

PRISCILLA: I might be your guest, but I'd be much obliged if you didn't use obscene language in my presence.

HARRIET: It's not the language that's obscene. It's the behavior it describes. Tell me something. What sort of mother did you have?

SALLY: Harriet!

HARRIET: I'm just asking.

PRISCILLA: It's okay, I'm happy to answer. She was a God-fearing woman and a wonderful mother.

HARRIET: Well, that's great. And how about your father? Where on the spectrum of human behavior does he fit in?

PRISCILLA: That's none of your business.

CAMERON: Prissy!

HARRIET: You're right. It isn't. But why didn't you just say "He was a God-fearing man and a wonderful father"?

PRISCILLA: He was God-fearing . . .

HARRIET: Well praise the Lord, but was he a wonderful father? Like Sally's, for instance.

PRISCILLA: I don't wish to put up with this any longer.

HARRIET: I think you've just answered my question.

PRISCILLA: I will not stay here and listen to this! (*She rises. To* CAMERON.) I'm going back to our hotel. (*To* SALLY.) Congratulations on a healthy baby. Even if he is the product of an immoral union. Goodbye.

(PRISCILLA *leaves.*)

CAMERON: I'd better go after her. (*He gets up and heads for the door, grabbing his and* PRISCILLA'S *coats on the way. To* SALLY.) I'll call you.

SALLY: Okay . . . look Cam, I'm sorry.

CAMERON: It's okay. She should never have come with me.

(CAMERON *goes.*)

HARRIET: Well, that was short.

SALLY (*loud and angry*): Why did you do that?

HARRIET: Don't. You'll wake the baby.

SALLY: You provoked her!

HARRIET: Nonsense.

SALLY: Jesus, Harriet! They're family!

HARRIET: What does that mean, "they're family"?
SALLY: It's not just semantics.
   (*The baby wakes up, making a protesting noise.*)
SALLY: Shh, shh, it's okay, it's okay . . .
   (SALLY *gets up, heading for the bedroom. Then she turns back to say:*)
   You know, they aren't having any luck trying for a baby. It would be nice if you could make some effort, show a little sensitivity.
HARRIET: All right, I'm sorry. Next time, I'll try to be more helpful.
SALLY: Thanks. (*Beat.*) Maybe you could . . . I don't know . . . talk to Cam as a doctor. Maybe there's something you can do for them.
HARRIET: You mean, offer him a consultation? You'd be okay with that?
SALLY: I just want us to help.
HARRIET: You think he'd go along with that?
   (SALLY *gets up and walks toward the bedroom.* HARRIET *sighs. She passes a hand across her face. She drinks some coffee and tries to relax. Her phone rings. Grabbing it and switching it off.*)
   Oh, shut up!
   (*End of scene 3.*)

Scene 4

   (SALLY *and* HARRIET*'s apartment, one month later.* HARRIET *is reading a medical journal. There is the sound of a baby moaning a little.* HARRIET *listens. Then it stops. She carries on reading.* SALLY *enters, wearing pajamas and looking tired.*)
HARRIET: Is he asleep? (SALLY *nods, then walks over to* HARRIET *and curls up beside her.*) That took forever.
SALLY: I think he's starting to teeth.
HARRIET: I doubt it, it's too early. Just give him some baby aspirin and be done with it.
SALLY: I am not giving my child drugs unless he really needs them.
HARRIET: And so, another sleepless night in store.
SALLY: I don't care. He's my son.
   (HARRIET *broods on this for a moment. There is a pause.*)
HARRIET: Glass of wine? A bite to eat?

SALLY: I think I'll just go to bed.

HARRIET: Massage?

(*This perks* SALLY *up a bit.*)

SALLY: Let me do it to you . . . I don't spend enough time babying you.

(HARRIET *lies on her stomach on the floor with* SALLY *straddling her.* SALLY *begins massaging* HARRIET*'s shoulders.*)

HARRIET: Hm . . . hmm . . . a little to the left . . . that's it . . . hmm . . .

SALLY: You've been working too hard.

(SALLY *continues massaging* HARRIET*'s shoulder. She sees one of* TUCKER*'s toys on the floor; she smiles and chuckles softly.*)

HARRIET: What's that chuckle for?

SALLY: Oh, nothing . . . just . . . he's so perfect. Like a sculpture.

HARRIET: I know. I love holding your little boy.

(*A beat.*)

SALLY: What do you mean, "your little boy"?

HARRIET: Nothing really . . . he's just . . . he's your little Tucker.

(*Another beat.*)

SALLY: What's wrong with "Tucker"?

HARRIET: Nothing. It's just a bit . . . Southern.

SALLY: So?

HARRIET: All I'm saying is, you picked the name.

SALLY: And I asked you . . . and you said you didn't mind.

HARRIET: You didn't ask if I had any preferences.

SALLY: Did you?

HARRIET: What's the difference? Once you proposed "Tucker," the only option you left me was my mother's maiden name.

SALLY: "Beppuchinsky Parker?" (*She laughs.*) Quite a mouthful.

HARRIET: You think "Tucker Parker" is euphonious?

SALLY: Why didn't you say that then?

HARRIET: My heart went out to you when you asked "would you mind 'Tucker'?" I saw how much baggage you were dumping on the table when you suggested your mother's maiden name.

SALLY: What is all this, Tucker versus Beppuchinsky?

HARRIET: It's not that.

SALLY: Then what? Come on.

HARRIET: Look, all I'm saying is, he's your flesh.
SALLY: How do you mean, he's my flesh?
HARRIET: I mean, the flesh comes from you. I am only a bystander.
SALLY: Harry!
HARRIET: What?
SALLY: That isn't fair.
HARRIET: It has nothing to do with fairness. I'm facing reality.
SALLY: Reality? What reality? Come on, out with it.
HARRIET: It isn't that easy.
SALLY: Honey! Do you want to tell me or don't you? Otherwise I'm going to bed.
HARRIET: No, wait. (*Pause.*) You know, soon he'll start talking. What will he call me? Surely not "Mother"?
SALLY: Why not?
HARRIET: Two mothers? In just a few years it would become damn confusing.
SALLY: All right, then. "Co-mother"?
HARRIET: Forget it.
SALLY: "Mother" and "Mummy."
HARRIET: Sure. And you know who'll be the "Mummy."
SALLY: Well . . . how about "Harry"?
HARRIET: You see? He'll call you "Mom" and me "Harry."
SALLY: Okay. I'll train him to call me "Sally."
HARRIET: I don't like it when children call their parents by their first names. They sound like strangers.
SALLY: Then we'll invent names we both can live with. Okay?
HARRIET: Okay. (*She stands up.*) Sally . . .
SALLY: Yes?
HARRIET: I don't know how to break it to you . . .
SALLY: Jesus, Harriet! Are you having an affair?
HARRIET: It's more serious than that.
SALLY: What can be more serious than having an affair?
HARRIET: I want a baby.
SALLY: We have a baby.
HARRIET: A child of my own.

SALLY: You've got a child. His name is Tucker.

HARRIET: Well of course. But it's not the same. I'm like a stepfather.

SALLY: Stepmother.

HARRIET: No, step*father*. Somehow, I seem to have acquired all the features of a father. Even your prissy sister-in-law noticed.

SALLY (*she steps back*): Are you jealous of Tucker?

HARRIET: In a way . . . in a very fundamental and stupid way . . . yes, I'm jealous. He's changing me from a partner into a stepfather. (*Beat.*) I used to think that the really fabulous thing about a relationship between two women is that everything is possible . . . not just in sex. *Everything* . . . including being a father *and* mother. But right now, I am mostly just a father.

SALLY: She really got to you, didn't she?

HARRIET: Priscilla? No! This has nothing to do with her.

SALLY: Are you sure?

HARRIET: Yes! It isn't Priscilla who is making me feel this way.

SALLY: It's me.

HARRIET: No! Yes! I don't know. It's *me*.

SALLY: How long has this been going on?

HARRIET: I don't remember . . . maybe since he started sharing the bed.

SALLY: You're jealous.

HARRIET: No!

SALLY: You are! But you've no reason to be.

HARRIET: That's easy for you to say. (*Beat.*) I've got to give it a try.

SALLY: A try? Trying to become a mother? And if you don't like it, after you've "given it a try," what then? You gonna hand it over to me: "Sorry, honey, I gave it a try, but it didn't work out. Let's go back to where we started . . . but with two kids."

HARRIET: It won't be like that.

SALLY: You're damn tootin' right it won't. Because the baby doesn't go back into the test tube if you don't like it anymore.

HARRIET: Sally! That's grossly unfair!

SALLY: Is it? Would you be willing to take care of *both* our children while I go back to work? Or you'll look after yours and I'll look after mine? I'm sure that if I ask for my job back, they won't mind if I bring Tucker into the studio with me. (*She imitates her news reader's voice.*) "And now back to the White House for some breaking news on . . ." (*She interrupts herself to imitate the sound of* TUCKER *crying.*) Ahhhhhh . . . harhhhhhhhhh! (*Beat.*) And then I'll explain to the listeners that this isn't a new baby in the White House but my son Tucker who is hungry. Of course, you'll have a job explaining to your patients in their underwear why you are changing your baby's diaper while discussing their incontinence problem.

HARRIET: Now you're being stupid. (*Beat.*) Of course, I've thought about this!

SALLY: I don't believe it! All you've thought about is, "I'll have what she's having." (*There is a simmering pause.*) I can't deal with this. I'm going to bed.

HARRIET: Fine!

(*The door slams as* SALLY *exits, leaving* HARRIET *alone and upset.*)
(*End of scene 4.*)

Scene 5

(*Lights on* HARRIET *and* CAMERON *in her office, next morning.* HARRIET *now wears a doctor's white coat.* CAMERON *waits while she studies some test results.*)

CAMERON: Mind if I ask a question?

HARRIET: Ask.

CAMERON: What got you into urology?

HARRIET (*laughs*): What makes you ask?

CAMERON (*embarrassed*): You know. A woman urologist and . . .

HARRIET: A lesbian?

CAMERON: Well . . . yes.

HARRIET: It wasn't for the obvious reasons . . . not the ones you and most other men . . . especially so-called straights . . . think.

CAMERON: I'm sorry, I shouldn't have asked.

HARRIET (*laughs*): That's okay, you can ask your sister-in-law anything you like. Some people call urologists medical plumbers, but just think what life would be without plumbers.

CAMERON: So you wanted to be a medical plumber?

HARRIET: Not just that. I'm also interested in male reproductive function.

CAMERON: Oh?

HARRIET: You don't approve?

CAMERON: On the contrary, I think it's neat. But what do you do there?

HARRIET: Well, I've done my share of vasectomies.

CAMERON: You call that "reproduction"?

HARRIET: Vasectomies prevent further reproduction. Lots of men find that important. (*He frowns.*) Although usually only after they've had children.

   (CAMERON *nods, apparently finding that more acceptable.*)

CAMERON: And . . . what about the other way around?

HARRIET: Men who have trouble reproducing? (*Embarrassed,* CAMERON *nods.*) They interest me even more.

   (*A pause.*)

CAMERON: You never asked why I came for such a short visit all the way to San Francisco.

HARRIET: Other than as a patient? (*She smiles to make him feel more comfortable.*) You missed your sister and your nephew.

CAMERON (*embarrassed*): I came to ask whether you could help us.

HARRIET: I read your results. Your sperm are fine. Perhaps the problem is Priscilla.

CAMERON: Yes, the doctor said she had obstructed fallopian tubes and ovarian cysts.

HARRIET: That's a tough call. (*Brief pause.*) Have you considered adoption?

CAMERON: Priscilla wants to give birth to our child.

HARRIET: *Our* child? Genetically, half of it would be yours if your sperm is used, but the other half has to come from an egg donor. And even then there's no guarantee the transferred embryo would implant in your wife's uterus.

CAMERON (*impatient*): I know all that.

HARRIET: Okay. Just checking you do.

CAMERON: Why did you agree to see me?

HARRIET: It was Sally's idea. . . . And I want to help if I can.

CAMERON: I see. Well, thank you. It's difficult. I mean, I've been losing sleep over this.

HARRIET: Don't talk to me about losing sleep.

CAMERON (*with a smile*): How's the little feller?

HARRIET: The little feller is doing fine. (*Beat.*) Whatever it is, you can tell me in confidence. It will never leave this room. (CAMERON *nods, appreciating that, thinking it over.*) You're considering IVF, is that it?

CAMERON (*after a moment*): What else can we do? We want a child . . . powerfully bad.

HARRIET: That bad, huh? Well, I am sorry to hear that.

CAMERON: You are?

HARRIET: Maybe I wouldn't choose Priscilla as a dinner guest, but . . . you're family to Sally . . . and therefore to me.

CAMERON: You believe in family?

HARRIET: That's part of what brought Sally and me together. You know, when your parents and you . . . well, anyway . . . it broke her heart. So . . . we decided to make a new family.

CAMERON: I never intended to turn my back on Sally. Mom and Dad . . . and then Priscilla . . .

HARRIET: Enough of that. (*For a moment,* HARRIET *lets* CAMERON *struggle to contain his feelings.*) I could help you.

CAMERON: Really?

HARRIET: I'm an urologist . . . I deal mostly with men . . . but I know lots of infertility specialists. I could help you find an egg donor.

CAMERON: You could? A suitable one?

HARRIET: What does "suitable" mean?

CAMERON: An egg that'll help get us someone like Tucker.

HARRIET: That's all?

CAMERON: He's a wonderful kid.

HARRIET: That he is. (*Beat.*) But you'd leave the choice of donor up to a woman you don't approve of?

CAMERON: You say you understand about family. And you're Tucker's . . . (CAMERON *is stumped.*)

HARRIET: Father?

CAMERON: Not that.

HARRIET: Well, what about "co-mother"?

CAMERON: Listen. Approval and trust are not the same. I don't approve of you . . .

Taboos | 71

HARRIET: But you trust me. You know, you could've gone to a specialist, someone anonymous.

CAMERON: They wouldn't understand. You know . . . you'd keep it in the family. That's what my parents taught me. If you keep it in the family, you can solve all your problems.

HARRIET: Unless the family *is* the problem, but we won't go into that. Okay. I'll do what I can for you.

CAMERON: I'm much obliged. (*Beat.*) What do I owe you?

HARRIET: Well . . . I'll think of something.

CAMERON: You bet.

HARRIET: Does Priscilla know about this meeting?

CAMERON: Not yet.

HARRIET: But you've discussed the issues?

CAMERON: No.

HARRIET: Well, sooner or later . . . she'll need to know what you've got in mind. And you'd better make it "sooner."

CAMERON: I guess so.

HARRIET: Okay. I'll leave that up to you.

CAMERON: Thanks. I truly appreciate it.

HARRIET: You're welcome. I think we're done here.

CAMERON (*getting up to leave.*) So what do I owe you for this?

HARRIET: Nothing right now. I'm sure I'll think of something. Bye for now.

CAMERON: Bye. (*He starts to leave, then suddenly stops and quickly and shyly kisses her on her cheek.*) I think you're neat, Harriet.

HARRIET: That's the nicest compliment I've had in some time.

(*Lights out as he exits.*)

(*End of scene 5.*)

Scene 6

(SALLY *and* HARRIET's *apartment. Evening, same day.* SALLY *is tidying up* TUCKER's *things.* HARRIET *comes in from work, walks past* SALLY *to the sofa, glancing at* SALLY *as she passes.* SALLY *goes to* HARRIET *and hugs her.*)

HARRIET: Hi.

SALLY: Hi. (*Pause.*) How was your day?

HARRIET: Fine. How was yours?

SALLY: Good. Well . . . I swear Tucker was trying to smile for the first time today.

HARRIET: Hmmm . . . that's nice.

 (*Beat.*)

SALLY: Honey . . . I'm really sorry about last night.

HARRIET: Me too.

 (*Pause.*)

SALLY: So you want a baby?

HARRIET (*with relief*): Yes, I do.

SALLY: Well, talk to me about it. Help me to understand what's going on in your head.

HARRIET: You really want to know? I'd like to start the procedure soon.

SALLY: How soon?

HARRIET: Well it's going to take at least two months' preparation because I'd like to use IVF.

SALLY: What about the sperm? Are you planning to go to a sperm bank?

HARRIET: I've already done that.

SALLY: What! And you didn't feel like telling me?

HARRIET: I realized a sperm bank was not for me. That's why I didn't mention it to you. I just need to see how the superovulation will go. I'm now thirty-seven. Not too old to have a baby provided I'm fertile . . . but old enough to take precautions.

SALLY: You mean amniocentesis?

HARRIET: No . . . I don't want to take that route. Then . . . if anything is wrong . . . abortion would be the only alternative, because I'd already be three months pregnant. I'm opting for pre-implantation genetic screening of the embryos.

SALLY: In other words, no turkey baster?

HARRIET: No turkey baster. No ordinary artificial insemination. It's got to be ICSI. A single sperm injected into my egg. That feels right. (*Beat.*) I've been thinking about it quite a bit.

SALLY: You sure have! Wow!

 (SALLY *cuddles up to* HARRIET.)

HARRIET: Yes.

SALLY: So in all your thinking have you worked out who the lucky donor might be?

HARRIET: Yes.

SALLY: Who?

HARRIET: Well . . . you know I saw Cam today.

SALLY: Him . . . as the donor?

HARRIET: Yes.

SALLY: Let me get my head around this. You are seriously thinking about using Cam as your sperm donor, with all his Christian baggage . . .

HARRIET: Christian baggage aside, he's a perfect candidate. His sperm seem totally healthy. But more importantly, you and Cam are from the same gene pool. I want to keep it in the family. Sally, I want to see your face reflected in my baby.

SALLY: Cam's sperm are healthy?

HARRIET: As far as I can tell.

SALLY: But Cam would never agree to be just a sperm donor. What about Priscilla?

HARRIET: Let's worry about that later. If I convince him . . . he might not even tell her. (*Pause.*) So . . .

SALLY: Harriet honey, you have to learn to talk to me about this stuff.

HARRIET: I know, I know absolutely. You're so right . . . so what do you think?

SALLY: I think we're going to have another baby.

(*End of scene 6.*)

Scene 7

(PRISCILLA *and* CAMERON's *home in Jackson, three months later.* CAMERON *is standing in the middle of the room. From an open door to the bedroom, a suitcase comes flying toward him. He ducks to avoid it.*)

CAMERON: Priscilla! (*Some shirts are hurled at him through the door.*) Honey! (*Some of his trousers are thrown at him.*) Prissy! You don't want me to leave.

PRISCILLA (*offstage, as some of his shoes are launched at him*): Adulterer!

CAMERON: What are you talking about? I wasn't even in the same room with her!

(PRISCILLA, *enraged, enters from the bedroom brandishing a pair of scissors and* CAMERON's *ties.*)

PRISCILLA: Your sperm committed adultery!

(PRISCILLA *cuts through one of his ties.*)

CAMERON: Now, hold on . . . those are my ties.

PRISCILLA: We're married . . . our property is community property. That includes your ties and your sperm . . . every single one of them. (*She cuts another one of his ties in half.*) The Holy Scripture says so.

CAMERON: Where does it say so? (*She cuts another tie, quickly.*) Never mind. What about when I discard it?

(PRISCILLA *stops.*)

PRISCILLA: When you what?

CAMERON: You know what I mean.

PRISCILLA: I don't know what you're talking about.

CAMERON: Sure you do.

PRISCILLA: I'm not a mind reader.

CAMERON: Well . . . (*He hesitates.*) . . . self-indulgence. That's not adultery.

PRISCILLA (*shocked*): You do what?

CAMERON: Masturbate.

(PRISCILLA *cuts another tie.*)

PRISCILLA: Cameron Parker. I don't ever want to hear that word in this house.

CAMERON: Okay! So! It's sinful to use the word, but not a sin when you practice it?

PRISCILLA: What's that supposed to mean?

CAMERON: I've seen you do it.

PRISCILLA: Once! And I begged the Lord to forgive me.

CAMERON: And you've never sinned that way again?

(PRISCILLA *can't answer that. Instead:*)

PRISCILLA: Let us pray. (*She sinks to her knees with her eyes closed.*) O Heavenly Father, purify our hearts and minds so that we are relieved of the temptations of lust. Strengthen not only me . . . (*She opens her eyes and notices that* CAMERON *is still standing and has not joined her in prayer.*

Before continuing, she first taps the floor with one of her hands, implying that she wishes him to kneel as well.) ... but also my husband against the malice and snares of the Devil. Let my husband's eyes be opened to his transgressions, and may he go forth and sin no more. Amen.

(*At this point,* CAMERON *yields by kneeling next to* PRISCILLA *and continues immediately.*)

CAMERON: Cleanse my heart, O Lord, each time I trespass. (*Beat.*) As I surely will again. And may the Holy Spirit protect me that I may continue to serve you faithfully in spite of my imperfections. (*Beat.*) And those of my wife. Amen.

PRISCILLA: How dare you even speak of self-indulgence? "Thou shalt not spill thy seed in vain." Genesis 38.

CAMERON: Onan wasted his seed by spilling it on the ground. Mine wasn't wasted. Mine may help Harriet conceive a child. Is that so sinful?

PRISCILLA: That's it! That is . . . it!

(PRISCILLA *grabs her cell phone.*)

CAMERON: What are you doing?

PRISCILLA: I'm calling your sister! (*Beat.*) What's the number?

CAMERON: Don't you think you should let Harriet tell her, in her own time?

PRISCILLA: Don't mention that woman's name to me! (*She throws the phone down in frustration.*) Why did it have to be you?

CAMERON: I might say it was the Christian thing to do . . . helping someone who was in need.

PRISCILLA: Shame on you! With your own sister-in-law!

CAMERON: Prissy! They aren't married!

PRISCILLA: Worse! Your own sister's lesbian lover! How could you?

(PRISCILLA *bursts into tears.* CAMERON *approaches her cautiously.*)

CAMERON: I wanted to know if I was fertile . . . whether it's also my fault we still have no baby. This way I could find out and, at the same time, keep it in the family . . . so to speak.

PRISCILLA: "Keep it in the family"?

CAMERON: So to speak.

PRISCILLA: That is . . . sick! (*Beat.*) It's my fault . . . I knew it all along. I'm paying for my sins.

CAMERON: Don't start with that. It's not your fault . . . you can't help it. The doctor says you have obstructed fallopian tubes and ovarian cysts.
PRISCILLA: Stop changing the subject!
CAMERON: Prissy, what's done is done. But you and me, that's different. Don't you see? I did it because I love you. Because I want us to have a baby. If I made a mistake, I'm sorry. But, at least, this way, we all know where the genes come from. And whatever you think, it's not incest. I didn't want to hurt you . . . I just believe in family . . . in keeping things in the family. You know that. Nothing's more important to me. That's why I want us to have a child . . . real bad. What's wrong with that? (*Beat.*) "And I will bless them that bless thee . . ."
PRISCILLA: "And curse them that curseth thee."
CAMERON: "And in thee shall all families of the earth be blessed."
PRISCILLA: Genesis.
CAMERON: Chapter 12, verse 3.

 (*Long pause.*)

PRISCILLA: How are we going to solve *my* problem?
CAMERON: Another woman could help.
PRISCILLA: Don't think I haven't thought about that.
CAMERON: Well?
PRISCILLA: But who would do that? I know, I wouldn't do it for another woman . . . if I were fertile.
CAMERON: Some generous woman might.
PRISCILLA: How would I find one? I'd be too embarrassed to ask.
CAMERON: You want me to try?
PRISCILLA: You would? (*Beat.*) But if you do, don't tell me about her.
CAMERON (*surprised*): You wouldn't want to know who she was? What she looked like?
PRISCILLA: Of course I'd want to know *something* about her: age . . . health . . . family background (*Beat.*) and, of course, religion.
CAMERON: Religion isn't genetic. It's just an egg.
PRISCILLA: Still . . . I'd be more comfortable if it were a Christian egg. But I wouldn't want to meet the donor . . . or see a photo. I guess just like in a sperm bank. You get lots of information . . . genetic history, color,

education . . . even hobbies and favorite authors . . . descriptive stuff . . . but no photo or name.

CAMERON (*astonished*): Hobbies? Favorite author? How do you know that? (*Beat.*) Pris! You didn't go to a sperm bank, did you? (*He looks at* PRISCILLA, *suddenly flabbergasted.*) You did?

PRISCILLA: I didn't *go* to a sperm bank . . .

CAMERON: But?

PRISCILLA: I *looked* at some . . . on the Internet. You'd be amazed what you can find there. More information on an anonymous sperm donor than I ever knew about my own husband.

CAMERON: Why didn't you tell me any of that?

PRISCILLA: I was scared.

CAMERON: Of me?

PRISCILLA: I don't know. Maybe I was scared . . . that I'd be tempted.

CAMERON: And now?

PRISCILLA: I guess I'm about to yield to temptation. But you've got to tell me . . .

CAMERON (*shocked, he interrupts her*): Wait! You just said you didn't want to know the identity of the egg donor.

PRISCILLA: I don't want to know about her, I just want you to tell me how you fertilized her egg. Promise you'll do it with ICSI . . . and no other way!

CAMERON (*flabbergasted*): ICSI? How do you know about ICSI?

PRISCILLA: I looked it up on the Internet. With ICSI only one sperm is misused.

CAMERON: Misused?

PRISCILLA: Because it isn't natural, but I wouldn't want you to do it the natural way . . . with millions of sperm. It seems less sinful . . . doing it with only one.

CAMERON: Well, we may need several sperm. With ICSI, one generally injects more than one egg.

PRISCILLA: And what will you do with the extra embryos?

CAMERON: Why do you need to know these things?

PRISCILLA: Because I do!

CAMERON: Get them to freeze them.

PRISCILLA: For how long?

CAMERON: I don't know. Until we're sure we won't need them anymore.

PRISCILLA: And then?

CAMERON (*getting impatient and flustered*): I don't know. Give them up for adoption?

PRISCILLA: Hush! I don't want to hear any more about it. (*She falls to her knees, dragging* CAMERON *down with her hand.*) O Lord, have pity on these two sinners who want a child so badly, and grant our wish for a successful birth. Whatever impure thoughts we had . . . whatever improprieties we committed . . .

CAMERON: . . . or may commit in the future . . .

PRISCILLA: . . . whatever secrets we kept from each other . . .

CAMERON: . . . or may keep from each other . . .

PRISCILLA (*startled, she quickly looks at him and then interjects*): . . . or discover unbeknownst to the other . . .

CAMERON: . . . or even those never uncovered . . .

PRISCILLA: . . . but especially . . .

CAMERON (*he quickly interrupts*): . . . forgive us, because we mean no ill.

PRISCILLA: Still . . .

CAMERON (*he quickly interrupts again, more forcefully*): Amen!

PRISCILLA: But . . .

CAMERON (*even louder and more forcefully*): I said, Amen!

    (CAMERON *rises from his knees and pulls* PRISCILLA *up with him.*)

    (*End of scene 7. End of act 1.*)

Act 2

Scene 8

    (SALLY *and* HARRIET*'s San Francisco apartment, four months later.* SALLY *is reading, while* HARRIET—*four months pregnant*—*sits quietly staring into space.* MAX *gently rocks baby* TUCKER *in his cradle.* CAMERON *looks fondly at the baby.*)

CAMERON: Cute kid. He hardly ever cries.

MAX: Why should he? He's got two mothers and me.

CAMERON: Three parents.

SALLY (*looking up from her book*): Cam . . . there are *two* parents . . . me and Harriet.

CAMERON: What about Max?

SALLY: He's the sperm donor.

CAMERON: That makes him the daddy.

SALLY: Biological daddy.

CAMERON: Jeez, Sally . . .

MAX: Forget it, Cam. I'm like you . . . you're the biological uncle . . . I'm the biological father. (*He rocks* TUCKER'S *cradle*.) At this stage, Tucker doesn't know the difference . . . and I'm content to leave it that way.

CAMERON (*shocked, he interrupts* MAX): Father and uncle aren't the same. I don't know whether I could handle it.

HARRIET (*who until now had been quietly concentrating, looking into space, gives a start*): You don't really mean that, do you?

CAMERON (*embarrassed*): I guess not.

HARRIET: I'm relieved. (*She walks over to* MAX *and gives him an affectionate shove.*) Move . . . sperm donor. It's time for one of the parents to take over. (*She takes the sleeping baby out of the cradle and rocks him gently in her arms as she sits down.*) Actually . . . why don't you and Sally take a walk. I want to talk to the uncle . . . in private.

SALLY: Without the co-mother?

HARRIET: Even without her.

MAX (*good-naturedly*): Come, Sally . . . let's follow the doctor's orders.

(MAX *bends over to give* HARRIET *a kiss on the forehead and then he and* SALLY *exit*.)

HARRIET (*with* TUCKER *in her arms, looks at* CAMERON *until the baby starts fidgeting*): You're returning to Mississippi today?

CAMERON: I've got to . . . I can't leave Priscilla alone. She doesn't go out much now. She felt pretty bad until recently. Morning sickness.

HARRIET: You said she's three months pregnant . . . she'll soon be past the worst.

CAMERON: Yeah . . . she feels better now.

HARRIET: Cam, we've got to talk. Seriously.

CAMERON (*taken aback*): About what?

HARRIET: Us.

CAMERON: Us? What do you mean?

HARRIET: I'm four months pregnant.

CAMERON: That's the chief reason I came. I wanted to know how you were.

HARRIET: Well, as you see, I'm coping very well. Besides, I thought you came to see your nephew. Here, hold him.

> (HARRIET *hands* TUCKER *to* CAMERON, *who takes the baby rather clumsily, whereupon increasingly loud baby screams are heard. He tries to rock* TUCKER, *still clumsily, without affecting the screaming.* HARRIET *picks up the baby and holds him closely while gently patting his back. The screaming subsides . . . then stops.*)

> You need practice with babies. (*Long pause.*) Why did you want to know how I am?

CAMERON: Because . . . (*He hesitates out of embarrassment.*) . . . you know why . . .

HARRIET (*sharply*): No, I don't.

CAMERON: Jeez.

HARRIET (*sharply*): Tell me. Why?

CAMERON (*he points at her stomach*): I got you pregnant . . . so I felt responsible . . .

HARRIET (*quickly, almost angrily*): Hold it, hold it! You didn't get me pregnant . . . and you're certainly not responsible. (*Beat.*) Okay, listen. Listen very carefully: We used a few sperm of yours . . . seven to be precise . . . for injection into seven of my eggs. You wanted to know whether you were fertile . . . and I agreed to help you find out with one of the embryos. I did that only because I wanted a baby of my own together with Sally. And since she's your sister, she's contributing to the baby's gene pool through your sperm. It was my decision . . . and it's my responsibility. You got out of the loop, once you masturbated.

CAMERON (*dry with touch of irony*): Much obliged Ma'am . . . for this clear explanation.

HARRIET: I'm not finished. The moment the embryo implanted in my uterus, I gave you the rest to do as you pleased. That was the only bargain between us. That your wife became pregnant with one of those embryos is your responsibility . . . not mine. (*Ever more intense.*) When this boy is born (*She points to her stomach.*) he will be *my* son. And when Priscilla gives birth that will be *your* son. Is that understood? Let's not confuse those two sons.

CAMERON: It's not as simple as that.

HARRIET: Oh, yes, it is.

CAMERON: And . . . if I have feelings?

HARRIET: You deal with them. But you and your Christian wife have no rights over my son.

CAMERON: Is that a threat?

HARRIET: I wouldn't dream of threatening you. I like you. I'm just giving you the facts.

(*There is a tense pause.*)

CAMERON: How do you know they'll both be sons?

HARRIET: Because I wanted mine to be a son.

CAMERON: But that's no guarantee. God decides what we get and we'll be grateful for whatever blessing He bestows.

HARRIET (*gentler*): Cam, I don't want to argue religion with you. This is biology. (*Beat.*) We used ICSI for the fertilization, right?

CAMERON: Right.

HARRIET: Injecting one sperm into each egg, right?

CAMERON: Right.

HARRIET: The sex of the child is always controlled by the sperm. A Y chromosome-bearing sperm leads to a boy, an X chromosome-bearing sperm to a girl. I'm sure you learned that in high school—

CAMERON: So what are you telling me?

HARRIET: That the technology . . . it's called flow cytometry . . . has now been developed to separate X- from Y-sperm—

CAMERON (*taken aback*): And you used separated sperm?

HARRIET: Yes.

CAMERON: And you didn't tell me?

HARRIET: That wasn't part of the bargain. You wanted to know whether you're fertile. I wanted to have a son, and you wanted to have a child with your wife. There weren't any more eggs of mine left for new ICSI injections. I was generous enough to give you the remaining embryos and all of those were potential males.

CAMERON: Jeez!

HARRIET: Cam, stop using that word. It's driving me crazy. And what's wrong with your having a boy?

CAMERON: Nothing.

HARRIET: You see!

CAMERON: But picking the sex of the child is so . . .

HARRIET: Don't tell me . . . unnatural.

CAMERON: Unnatural, yes.

HARRIET: And you think ICSI is natural? Most of modern medicine is full of interventions and materials that cannot be found in nature. You think "unnatural" is automatically "unethical"? (CAMERON *falls silent.*) But I wanted to talk about something else.

CAMERON: You mean there's more?

HARRIET: You bet there is.

CAMERON: What is it?

HARRIET: I wish you and Priscilla had waited until Jan was born before starting with the embryo transfer.

CAMERON: Jan?

HARRIET: My son.

CAMERON: It that what you're calling him?

HARRIET: Yes . . .

CAMERON: Why did you want *us* to wait?

HARRIET: Wouldn't it have been prudent to see first if everything was okay with Jan? After all, the other embryos came from the same woman and the same man.

CAMERON: Any other reason?

HARRIET: Now . . . if everything goes according to schedule . . . the two boys will be born less than four weeks apart.

CAMERON: So what?

HARRIET: Biologically, they could be twins.

(*Pause.*)

CAMERON: I'm sorry. I don't get it.

HARRIET: Of course, not identical twins. But what are twins but siblings from the same biological mother and father . . . but born at the same time?

CAMERON: Your baby and ours won't be born at the same time.

HARRIET: If fraternal twins grow in separate sacs in the mother's uterus and one breaks prematurely, you can have twins born at somewhat different times . . . even some weeks apart. They'll still be twins.

CAMERON: Jeez!

HARRIET: Exactly . . . Jeez! If they're born that close together, they're essentially twins . . . and that, in turn, will lead to a very special bonding with each other . . .

CAMERON (*shocked*): What?

HARRIET: Whereas they would not be twins if they were born nine months or more apart. They'd be ordinary brothers. Now I hope you understand why you complicated matters by not waiting. (CAMERON *nods, reluctantly.*) Of course, only you and I will know that our sons might be twins since for everyone else, two different women gave birth to them two thousand miles apart. You said you would not tell your wife where the embryos came from.

CAMERON: She didn't want to know.

HARRIET: What if she changes her mind?

CAMERON: Jeez, Harriet . . .

HARRIET: No "Jeez." I need to know what you will do if she changes her mind.

CAMERON: I don't know what I'll do!

(*Pause.*)

HARRIET: Well . . . whatever you decide, don't you ever forget: My son belongs to Sally and me. Your son is yours and Priscilla's. Is that understood?

CAMERON: I . . . I don't know.

(*End of scene 8.*)

Scene 9

(*A few days later. A room in* PRISCILLA *and* CAMERON*'s home in Jackson.* PRISCILLA, *in T-shirt and loose pants, lies flat on the dining table in a position that would prevent her seeing a person entering through the door behind her. Around her waist is a tightly wound wrapping* *from which an electric cord protrudes. The cord is connected to a CD player that is sitting on her stomach. She uses her left hand for "conducting" and the right for stopping and starting the CD player. Heard is some segment from Mozart's Oboe Concerto in C major, K. 314—low enough so that her words, which she enunciates carefully in recitative style, quoting from the Psalms, chapter 1, can be understood clearly.*

Unbeknownst to PRISCILLA, CAMERON *enters, wearing a jacket and carrying a small suitcase. He stops, surprised to see his wife on the table, but says nothing.*)

PRISCILLA: "Blessed is the man that walketh not in the counsel of the ungodly, nor standeth in the way of sinners, nor sitteth in the seat of the scornful . . ."

(*At this stage* PRISCILLA *shifts into an ordinary loud voice in a rapid, admonishing tone without shutting off the music.*)

Ashley, honey, that means you've got to be good and stay good . . . and "good" means that you must not sit in the "seat of the scornful." Remember that, Ashley: never sit there! Never!

(*She resumes quoting in her earlier biblical recitative style from the Psalms, chapter 4, while conducting in the air with her left hand.*)

"Stand in awe, and sin not. Commune with your own heart upon your bed, and be still."

(*She touches her stomach as if she were feeling some pain, then switches to her ordinary loud voice in an admonishing tone without shutting off the music.*)

Ouch! You're rumbling around too much, Ashley, honey. You must be still in mummy's tummy; it says so in the Psalms. So pay attention to your mother . . . and of course the Lord.

(*She resumes quoting in her earlier biblical recitative style from the Psalms, chapter 8, while conducting in the air with her left hand.*)

"Out of the mouth of babes and sucklings . . ."

(*She quickly switches to loud conventional speech.*)

Don't forget . . . in just a few months that'll be you, Ashley . . . and "suckling" means you'll be drinking your mummy's milk.

CAMERON (*steps slightly forward, but is still not noticed by* PRISCILLA): Jeez, Pris! What's going on?

(PRISCILLA *does not hear him. She resumes quoting in her earlier biblical recitative style from the Psalms, chapter 16, while conducting in the air with her left hand.*)

PRISCILLA: "Therefore my heart is glad, and my glory rejoiceth: my flesh also shall rest in hope . . ."

CAMERON (*steps farther forward and interrupts in a loud voice*): "For thou wilt not leave my soul in hell; neither wilt thou suffer thine Holy One to see corruption."

(PRISCILLA *jumps up and the CD player falls to the floor, cutting off the music.*)

PRISCILLA: My God, you scared me! What did you say?

CAMERON: The end of chapter 16 in the Psalms. But what are *you* doing?

PRISCILLA: Prenatal training of Ashley.

CAMERON: And who's that?

PRISCILLA: Our son.

CAMERON: How do you know we'll have a son?

PRISCILLA: I can feel it. But if the Lord provides otherwise, Ashley will also work for a girl. It's a good Southern name.

CAMERON: But we agreed we'd wait until the baby is born before picking a name. (*Beat.*) Together!

PRISCILLA: I've got to call him something while I talk to him.

(PRISCILLA *pats her stomach.*)

CAMERON: And you never told me?

PRISCILLA: I was planning to . . . but you've been away.

(PRISCILLA *points to his suitcase.*)

CAMERON: What prenatal training? Music?

(CAMERON *bends down to pick up the CD player, which suddenly resumes the Mozart.*)

PRISCILLA (*leans over to shut off the music*): The Mozart Effect. Listening to Mozart raises people's I.Q. It's called the transformational power of music.

CAMERON: How do you know about that?

PRISCILLA: It says so on the Internet. Babies are smarter after listening to Mozart. And rats exposed to lots of Mozart made fewer errors running in a maze.

CAMERON: I don't know about rats and Mozart, but you're spending too much time on the computer. It isn't good for him.

(CAMERON *points to her stomach.*)

PRISCILLA: Too much radiation?

CAMERON: Too much information.

PRISCILLA: For prenatal Christian training you can't start early enough. If there's a Mozart Effect, why not also a Jesus Effect? Listening to Scripture can't hurt little Ashley. (*She pats her stomach.*) That's why I was quoting from the Psalms.

CAMERON (*losing his calm, almost shouting*): Prissy! Listen! The baby . . .

PRISCILLA (*interrupts*): It's Ashley.

CAMERON: All right . . . Ashley. But Ashley can't hear these words . . . or your music! He isn't wired yet for that . . . not three months after the embryo transfer!

PRISCILLA: Well, that's why I have this wiring around me . . . so Ashley can hear it inside me. (*She points to the wrapping around her waist.*) Anyway, what do you know about baby wiring?

CAMERON: It takes much longer. Take Einstein. He didn't even start speaking until after he was two.

PRISCILLA: I only want a smart baby . . . not a little Einstein.

CAMERON (*shakes his head*): Jeez, Pris.

PRISCILLA: What's wrong with you? I'm not doing anything strange.

CAMERON: I'm sorry.

PRISCILLA: It's *my* prenatal training! Just let me do it!

CAMERON: All right, don't get hysterical . . .

PRISCILLA: Who's being hysterical?

CAMERON: Okay, I apologize. I spoke out of turn. Forgive me.

PRISCILLA (*after a pause*): You're forgiven.

(*She's about to go back into it when:*)

CAMERON: There's something we should talk about.

(PRISCILLA *stops.*)

PRISCILLA: Okay. You tell me.

(*A pause as* CAMERON *tries to find the words.*)

CAMERON: What . . . I mean, why were you lying on the table?

PRISCILLA: A hard surface is good for my back.

CAMERON: We could buy a harder mattress.

PRISCILLA: I don't mind the table. (*Beat.*) Cam . . . why did you have to go to California again?

CAMERON: I wanted to see how Sally is managing with Tucker. After all, he's my nephew.

PRISCILLA: That wasn't the only reason, was it?

CAMERON (*nervous*): What do you mean?

PRISCILLA: Well, you've clearly got something on your mind.

CAMERON: It can wait.

PRISCILLA: You just said we should talk.

CAMERON: I don't think now is the time.

PRISCILLA: It's Ashley, isn't it? Don't you like the name?

CAMERON: The name's great.

PRISCILLA: Well, what then?

CAMERON: I really don't think, in your present state of mind . . . I mean . . .

PRISCILLA: Did you find something out?

CAMERON: Well . . . kind of.

PRISCILLA (*nods*): About the donor?

CAMERON: Kind of.

PRISCILLA: I don't want to know. Whoever she is, I don't want to know. It's my baby now. The Lord will look after my baby.

(*End of scene 9.*)

Scene 10

(PRISCILLA *and* CAMERON's *living room in Mississippi, five months after scene 9.* PRISCILLA, *looking exhausted and disheveled, paces up and down with her hands over her ears. In the background through the open door are heard the periodic piercing screams of a three-week-old baby. This screaming at different decibel levels and intervals must continue through most of scene 10.*)

PRISCILLA (*almost screaming*): I can't stand it anymore! Three weeks . . . and he hasn't stopped!

CAMERON (*enters with the screaming baby in his arms*): Pris . . . try just once more.

PRISCILLA (*almost desperate*): I can't . . . I won't . . . you know it doesn't help.

CAMERON: It's only colic . . . you heard the doctor. It's acid reflux. His guts can't cope yet with your milk.

PRISCILLA (*again puts her hands over her ears*): I've had it with his screaming.

CAMERON: Here . . . hold him. I'll bring the new milk the doc prescribed. Maybe it'll help.

PRISCILLA (*takes the baby, but holds it in a manner that indicates little affection*): All right . . . but hurry! (*The baby screams.*) Ashley . . . will you shut up . . . please! *Please* shut up! (*New baby scream.*) Wait! *Please* wait! Daddy is bringing some new milk. (*New baby scream.*) Ashley! It's different from Mama's milk . . . I know you hate mine. But I can't help it, Ashley . . . it's all I got. (*Less loud baby scream.*) You should have waited, Ashley . . . you came four weeks too early. (*New louder baby scream.*) Your tummy hasn't learned yet to handle Mama's milk. (*New baby scream, whereupon* PRISCILLA *screams.*) Cam! Where are you with the damn milk? He's driving me crazy.

CAMERON (*rushes in with the bottle*): Prissy . . . don't use such language. . . . Here's the bottle.

PRISCILLA (*grabs the bottle, but quickly thrusts it back*): Damn you, Cam, damn you! It's too hot! Can't you do anything right?

(CAMERON *leaves. Shortly thereafter, the sound of running water is heard, followed by the baby's renewed scream.*)

Ashley, you got to wait! Your mama begs you. (*New louder baby scream.*) Ashley . . . I'll strangle you if you don't (*Very loudly.*) shut up! (*A few seconds of silence.*) Thank God!

(*There is a renewed loud scream, whereupon* PRISCILLA *roughly puts the baby into his cradle and falls on her knees, rocking back and forth with her hands clasped over her ears.*)

My Lord and Savior . . . I beg you on my knees to listen to this sinner. I do not deserve this punishment. I . . . do . . . not . . . deserve it!

CAMERON (*rushes in with the bottle, then stops, startled to find his wife on her knees, rocking back and forth*): What happened?

(*New baby scream.*)

PRISCILLA (*points to the cradle*): I can't handle it anymore.

CAMERON: Let me try.

(CAMERON *bends over the baby, offering the bottle, which produces prolonged silence.*)

PRISCILLA (*rises, surprised.*) Oh, Lord . . . you have listened to my prayer.

CAMERON: Maybe it was only your milk.

PRISCILLA: Let me do it.

(PRISCILLA *lifts the baby out of the cradle while continuing to feed it with the bottle. After a pause, the screams resume, ever louder. She tries to adjust the position of the baby and bottle without success. In desperation, she flings the bottle across the room and screams at the baby as she practically throws him into the cradle.*)

I hate you! I hate you! God only knows how I hate you!

(CAMERON *interrupts her, shouting almost as loudly, while picking up the cradle and heading for the door.*)

CAMERON: Pris! For heaven's sake! You don't hate him. You can't hate him. You mustn't hate him.

(CAMERON *takes the cradle into the other room.*)

PRISCILLA (*wailing*): Lord, forgive me. I didn't mean what I said!

(CAMERON *rushes in after carefully closing the door, thus dampening the noise of the periodic baby screams that will be heard in the background through the remainder of the scene.*)

CAMERON: Prissy, Prissy, Prissy! (*He takes her in his arms.*) You've got to pull yourself together.

PRISCILLA (*sobbing on his shoulder*): I know, hon . . . but I haven't slept since I came back from the hospital. There must be something wrong with him. I even tried Mozart.

CAMERON (*tries to cheer her up*): Maybe Ashley is overdosed with Mozart. But the good Lord . . . and time will help. It's his digestive system . . . I'm sure of it. The doctor said so. It's not uncommon with preemies.

PRISCILLA: I can't continue . . . I'm afraid what I might do.

CAMERON: We'll get some help.

PRISCILLA: How?

CAMERON: Last night, I called Sally.

PRISCILLA: Why not your parents?

CAMERON: I called them earlier.

PRISCILLA: And?

CAMERON: They said "pray." And when I told them you'd been doing that all the time, Dad just said "pray more." Well, I've done that and now I need some real, practical help and they're too old for that. So I phoned Harriet . . . I mean, Sally . . .

PRISCILLA: You phoned Harriet? Why did you do that? No lies to me, Cameron, no lies! You tell me!

CAMERON: No, I just . . . because her baby . . .

PRISCILLA: What? What about her baby?

CAMERON: He was born on the same day as Ashley . . .

PRISCILLA: And you didn't tell me that earlier?

CAMERON: I didn't think you'd take to it kindly.

PRISCILLA: But why her?

CAMERON: Look, all I wanted was to know how her son was. I wanted to know if they were having any problems . . .

PRISCILLA: Because you're also his father?

CAMERON: Jeez, Pris, I was the sperm donor!

PRISCILLA: But genetically . . .

CAMERON: Well . . . okay, yes . . . the father. That's why I called.

PRISCILLA: And?

CAMERON: And . . . she said he didn't make a peep. Drinks his mother's milk and sleeps.

PRISCILLA (*explosively*): God damn them! All of them!

(*During the following, the baby's screams get louder and louder.*)

CAMERON (*utterly shocked*): Priscilla Parker! You can't say that! (*He closes his eyes and lowers his head.*) Lord and Savior. Forgive my wife, who is not herself. Who is sick in mind and soul for reasons that only you understand. Let her examine herself to see whether she's still in the faith and do not punish her for blasphemies that she couldn't have meant. And since only thy words are true, shed light on my wife to lead her back to the path she has lost. Amen.

(*As* CAMERON *turns to* PRISCILLA, *he sees that she has picked up one of one of her dolls, her favorite one, and is rocking violently back and forth clinging on to it.*)

PRISCILLA: Make it stop . . . make it stop . . .

CAMERON: Come on . . . I'm getting you out of here. You need help.
(CAMERON *takes* PRISCILLA *by her hand and starts exiting as lights dim.*)
(*End of scene 10.*)

Scene 11

(*San Francisco apartment of* SALLY *and* HARRIET, *four weeks later.* MAX *gently rocks a cradle with* ASHLEY. HARRIET *sits across from him, breast-feeding* JAN. *Faint but distinct Mozart music—Oboe Concerto in C major, K. 314—is heard in the background.*)

MAX: I feel sorry for Cam. He's a good egg.

HARRIET: That he is.

MAX: I'm glad Sally is taking him to the airport. He needs cheering up before he gets back to Mississippi.

HARRIET: I'm wondering whether they're not releasing her too early. Recovery from postpartum psychosis can take time.
(*Incipient slight crying sounds from* ASHLEY *in cradle.* MAX *takes him out of the cradle and gestures to* HARRIET.)

MAX: They better trade places. I think Ashley is hungry . . . or maybe it's sibling jealousy.
(HARRIET *hands over* JAN *and takes* ASHLEY, *who immediately quiets down as he is fed.*)

HARRIET: Lucky he's got two aunts here. There's no way Cam could have managed by himself.

MAX (*wryly*): And four weeks later, here we are.

HARRIET: How d'you mean?

MAX: I mean, lots of things can change in four weeks.

HARRIET: Like Ashley getting over his colic?

MAX: Or the aunt not just being a wet nurse.

HARRIET: What are you trying to say, Max?
(HARRIET *places* ASHLEY *in the same cradle as* JAN, *then returns to her chair.*)

MAX: It's nice to see these two kids in the same cradle. Remarkable resemblance.

HARRIET: Many babies look alike at that age.

MAX: Bull shit.

HARRIET: What's that supposed to mean?

MAX: That it's time to face the music. I know you well enough that if you don't want to talk, you clam up. And with me you can do that. But it won't work with Sally . . . not much longer. And certainly not once Priscilla shows up here. (*Beat.*) Does Sally know how Priscilla got pregnant . . . and around the same time as you?

HARRIET: You should ask Ashley's parents.

MAX: I'm asking you. Or have you been carrying some sort of tranquilizer in your boobs since you've started nursing Ashley?

HARRIET: Give me a break. I thought you worked in the public defender's office, not the district attorney's.

MAX: Listen. I'm not just your brother. I'm Tucker's biological father, and I'm Jan's uncle. All of which puts me on your side. But am I now also an uncle of Ashley? That wasn't part of the deal.

HARRIET: What sort of deal are you talking about?

MAX (*good-naturedly*): To accept your co-parenting with Sally and not to interfere beyond certain limits. In other words, be an affectionate sperm donor and uncle . . . but nothing more . . . which suits me just fine.

HARRIET: Max . . . you've been wonderful.

MAX: I know . . . but now . . . I suspect we have a situation. So why have you clammed up?

HARRIET (*after a pause*): I'm afraid.

MAX (*sits down next to her with his arm around her shoulder*): You've got good reason to be. So tell me . . . was it one of your excess embryos? (HARRIET *nods her head.*) And you never told Sally?

HARRIET: No. That's why I'm scared.

MAX: What about Priscilla?

HARRIET (*shakes her head*): She didn't want to know about the egg donor. And Cam didn't want me to tell anyone else.

MAX: Taking Cameron's side rather than your brother's or your partner's?

HARRIET: I wasn't taking sides. I didn't want to get involved.

MAX: You think Sally will buy that?

HARRIET: Please, Max, don't continue. I feel terrible enough.

MAX: Does Priscilla know who *your* sperm donor was?

HARRIET: I think so.

MAX (*shakes his head*): As Cam would say, "Jeez, Harriet." (*Beat.*) I think you need a lawyer.

HARRIET (*sharply*): What kind of a lawyer?

MAX (*looks at her for a while, finally squeezing her shoulder*): Me.

HARRIET (*quickly*): But you don't know anything about family law.

MAX: That's not important. Your kind of family law hasn't been written yet. For a start, how would a lawyer define Cameron's roles? He's the father of twins . . . each from a different legal mother . . . some sort of uncle-in-law of his own son and uncle of his son's stepbrother . . . (*He throws up his hands.*) I could go on.

HARRIET: What's your point?

MAX: You need a lawyer . . . who'll keep you out of the clutches of other lawyers . . . and charge nothing for those services. Can you think of anyone else?

HARRIET: Who's going to sue me? Cameron?

MAX: Not on his own. He isn't that sort of guy. But Priscilla? She may be livid and jealous and revengeful . . .

HARRIET: For what?

MAX: Believing she'd been manipulated into giving birth to a child of a lesbian mother. How's that for starters? You know how she feels about lesbian couples. What if she sues for joint custody of *your* son . . . who is also some sort of stepson of hers? Or maybe exclusive custody so he's not being brought up in a heathen home? And sues you in a Mississippi court rather than here in San Francisco? God only knows how a Mississippi family court might rule.

HARRIET: You're out of your mind! She can't do that.

MAX: I hope you're right. But why, for heaven's sake, didn't you have some legally binding agreement with Cameron?

HARRIET: For what? For asking him to lend me a few sperm for injections into *my own eggs*?

MAX: "Lending"? It was an irrevocable transfer of title to property . . . property that you made much more valuable as a consequence of the use to which you put it.

HARRIET (*angrily*): There was no agreement about that property being returned upon request! And when I gave him the rest of the embryos, it was a gift . . . an unrestricted gift. I didn't even want to know what he'd use them for.

MAX: How could you *not* want to know?

HARRIET (*increasingly angry, bordering on guilt*): For me, an embryo in a Petri dish or in a freezer is an abstraction . . . a clump of eight or sixteen or thirty-two cells . . . nothing more. It's only when that abstraction is transferred into a woman and implants are we dealing with reality. And I was focusing on my own uterus . . .

MAX: Some legal journal will have a field day reporting this if it ever comes to trial.

(*A pause.*)

HARRIET: Oh, shit.

MAX: I can help you avoid this.

HARRIET: How?

MAX: She'll never sue without convincing her husband to join her.

HARRIET: And Cam will never do that.

MAX: I've never believed in the word "never," but I am a believer in preemption. How about getting everyone to agree to compulsory arbitration for whatever problems may arise from this spectacularly complicated reproductive mess between five adults . . . and not all of them consenting ones.

HARRIET: And if she doesn't go for it?

MAX: Then at least agree to some nonbinding one . . . or even just group counseling . . . with all of you.

HARRIET: And you.

MAX: Why me? I'm unlikely to pose you a problem.

HARRIET: Other than Jesus Christ . . . can you think of a candidate for that noncompulsory arbitrator of yours?

MAX: No, but I will.

HARRIET: Max, why are you so good to me?

MAX: What else would I do with my free time? You're lucky I'm still a bachelor.

HARRIET: You'll make a spectacular husband to some lucky woman.

MAX: I know . . . but I've become choosier since hanging around with you two . . . and Tucker.

(SALLY *enters, throws her coat on a chair and drops into another one.*)

SALLY (*groans*): I need a drink . . . after that scene at the airport.

MAX (*jumps up*): Yes, Ma'am.

SALLY: A stiff one.

MAX: Coming up.

(MAX *goes to the sideboard and pours her a whiskey.*)

SALLY: Cam's been told he can pick up Priscilla and bring her home. He sees it as great news through his rose-tinted glasses. I had to tell him it was a pigment of his imagination.

MAX (*laughs*): That's a good one. I may use that someday in court. (*Mimics formal tone.*) "Your honor . . . this is just some pigment of my learned counsel's imagination."

HARRIET: It's not funny . . . not now. What else happened?

SALLY: He thinks she'll want to see Ashley right away.

HARRIET: And one of us is supposed to bring the kid to Mississippi?

SALLY: No. Cameron will bring Priscilla here.

MAX: Wow!

SALLY: What do you mean by that?

MAX (*with a quick look to* HARRIET): Nothing.

HARRIET (*points to the cradle.*) Look at the two . . . sleeping so peacefully.

SALLY (*regret in her voice*): That won't last much longer.

HARRIET: I don't know how I'll cope with that.

SALLY (*sympathetically*): I know it's tough . . . after breast-feeding him for a month . . . and no more colic. We'll both miss him. But . . . you'll be going back to work . . .

(HARRIET *indicates with a look to* MAX *for him to leave the room.*)

MAX: Let me take the kids into the bedroom.

SALLY: Why?

MAX: I think the two of you need privacy.

(MAX *leaves with the two kids.* SALLY, *a little puzzled, watches him go.*)

SALLY: What was that about?

HARRIET: Sally. You noticed any similarity between Ashley and Jan?

SALLY (*shrugs her shoulders*): Some . . . but at that age many babies look pretty similar . . . especially if they share the same father.

HARRIET: It's not just that. I mean, common genes . . . perhaps more common than you think . . .

SALLY: More common than just from the same father? What are you driving at?

HARRIET: I don't know how to tell you this . . .

SALLY (*now it is sinking in*): Oh, my God, Harriet!

HARRIET: I am so sorry.

SALLY (*short silence*): They're twins? I thought we had a plan . . . two children . . . Tucker and Jan . . . with us their parents. Only to find out that was not enough for you?

HARRIET: Of course it was enough. How could I've known it would turn out like this?

SALLY (*almost screaming*): Well you're an idiot! A total, total, total idiot!

HARRIET: Listen! You have no idea what I've been through this last month. All my energy has gone into learning how to become a mother of one baby, while distancing myself from another.

SALLY (*still screaming*): That's not what I'm talking about! It seems you gave one of your embryos to Priscilla! Anything else I should know about? Is there a triplet hiding somewhere? Or sextuplets? After all, you had seven eggs. What were you aiming for? An ice hockey team?

HARRIET: Sally! Please!

SALLY: Please what?

HARRIET: I thought I was just doing an infertile couple a favor . . . until I realized I was nursing *my* child . . . a child that will be taken away from me.

SALLY: Oh, shit, Harriet! Shit, shit, shit!

(*A pause.*)

HARRIET: First, you must know that I love you . . . deeply. Before you walk away, you've got to hear that: *I . . . love . . . you.* I've loved you from day one.

SALLY: Oh, God. How did this happen?

HARRIET: Remember when we decided that Cam could be my sperm donor? And you were worried that he wouldn't want to be just a sperm donor?

SALLY: Get to the point.

HARRIET: When Cam told me he was looking for a surrogate embryo for Priscilla, but that she didn't want to know anything about the egg donor, I figured if I used his sperm and my eggs and gave him the excess embryos, he wouldn't bug me once he had his own baby . . . in Mississippi. About as far away from here as possible.

SALLY (*angrily*): But why didn't you tell *me* that was the deal? Is there something about me that makes you think you can't talk to me about these things? When I first met you I thought we had an understanding. You were my rock . . . solid, stable, reliable . . . that's why I picked you. But now everything's changed. I don't know who you are anymore. You're not my Harriet.

HARRIET: Sally! He asked me to promise not to tell anybody.

SALLY: And you think that promise is more important than our relationship? Jesus, Harriet! You had no right to *make that* promise.

(SALLY *picks up the coat she threw on the chair earlier, puts it on, and leaves, slamming the door behind her.* HARRIET *stands staring at the door.* MAX *appears in the bedroom doorway and is about to say something when we hear the sound of the key in the door. Still steaming.*)

The point is, you shouldn't have agreed to use any of your embryos without discussing it with me. It's that simple.

HARRIET: I could have done that . . . I should have done that . . . I wish I'd done that . . . but I didn't. (*Beat.*) I now know it was a mistake . . . a monumental mistake, but what more can I say?

SALLY: It's bad enough what you did. But twins? I am Jan's co-mother!

HARRIET: How was I supposed to know they would use my embryos so soon? And that it would implant at the first try? And that their baby would be born prematurely? If they had waited a few months . . . Ashley and Jan would just have been two children born two thousand miles apart from two different mothers. Instead they're sleeping in the same cradle . . . two twins of mine.

SALLY: Well, if you'd have talked to me. If Cam had talked to me . . . your partner in crime . . . all this might have come out before the first sperm was injected into your eggs.

HARRIET: Sally . . . I can't undo the biggest mistake of my life. (*Beat.*) I'm scared.

SALLY: Of what?

HARRIET: Of losing you. Of what I've done. Of Priscilla. Of what she might do. Once she finds out it was my egg, there's no way she's going to let me anywhere near him! You saw what she's like: she hates me! She thinks I'm an evil lesbian bitch!

SALLY: What did you expect? You weren't exactly friendly yourself.

HARRIET: Sally, you've got to help me. Poor Ashley will be brought up by a half-crazy, poisonous, Bible-thumping witch, who sooner or later will hate him from the bottom of her soul because he came out of my egg. She will punish him and it's all my fault.

SALLY: Harry! Don't get hysterical. The problem isn't just Ashley. If we get into arguments over who has what rights . . .

HARRIET: It's a nightmare.

(SALLY *exits to the kitchen. She comes back with a drink.*)

SALLY: Okay. This is what we'll do. I'll fly to Mississippi and break the news to Priscilla and try to contain the situation.

HARRIET: You will? Oh, you jewel, you wonderful Sally. (*Pause.*) I've made such a mess of things. (*Beat.*) Do you still love me?

SALLY (*a pause, she then grins*): No, I hate you.

(SALLY *embraces* HARRIET.)

(*End of scene 11.*)

Scene 12

(*A few days later. Lights on* HARRIET, SALLY, *and* MAX *in* SALLY *and* HARRIET*'s apartment.* SALLY, *still in her travel outfit, has her suitcase by her side.*)

SALLY: They'll be here soon. I told them to check into their hotel first.

HARRIET: I can't stand the suspense. How did it go?

SALLY: I think it's going to be all right. But you've got to promise me you're going to keep a lid on it when they arrive. For now we'll have to let Ashley go back with Priscilla. But that's not the end of it. . . . It's just stage one. Priscilla's still very fragile, so we have to be super careful.

MAX: How did you manage to tell Priscilla about the egg donor?

SALLY: Very gently . . .

MAX: And Cameron?

SALLY: He was listening.

HARRIET (*after a beat*): And how did she take it?

SALLY: She cried . . . she screamed . . . she threw her tea cup at him . . .

MAX: I hope it was empty.

SALLY: And then they prayed.

HARRIET (*with heavy irony*): You don't say.

MAX: And it worked?

SALLY: They have this thing called Forgiveness Time.

MAX: How did you fit into that?

SALLY: She invited me to join in, but I could see that Forgiveness Time was already going to be quite a long session without my getting involved, so I just waited outside.

HARRIET: Poor Cameron.

SALLY: She isn't that bad. (*Beat.*) She's had quite a lot to deal with. (*She suddenly remembers* TUCKER.) I'm a terrible mother! I haven't even kissed Tucker hello yet.

HARRIET: He's fine, he's fine. Max read him *Heather Has Two Mommies* . . . and now he's taking a nap.

SALLY (*guffawing*): *Heather Has Two Mommies!* I know I have a smart son, but this is for three- or four-year-old kids.

HARRIET: It's never too early to teach kids that two mommies is okay. Call it gay imprinting. (*The doorbell rings.*) My God! That'll be them. Max, would you?

(MAX *springs up from his chair and goes out. He re-enters with* CAMERON, *who is carrying a suitcase, followed by* PRISCILLA. SALLY *starts to rise.*)

CAMERON: Don't get up. (*He deposits the suitcase on the floor as* PRISCILLA *and* MAX *come farther into the room.*) Well. Here we are.

(SALLY *looks at* HARRIET. HARRIET *looks at* PRISCILLA.)

PRISCILLA: We've come to pick up Ashley.

HARRIET: I know.

(*Slowly,* HARRIET *gets up and hands* ASHLEY *over to* PRISCILLA, *who at once begins nuzzling him.*)

CAMERON: Okay, then.

HARRIET (*sharp tone*): You mean, that's all? Not even a "thank you"?

PRISCILLA: Oh . . . I thanked Sally back home. I sure appreciated what y'all did . . . taking care of him while I was ailing. But I can handle him just fine now . . . no more colic. I told Sally, we'll stay one more day and then (*She again nuzzles the baby.*) Ashley flies home with his mommy and sleeps in his own crib, won't you? You'll never know you were away from your mommy and your daddy.

HARRIET (*almost livid*): What about me? What role do you think I play in all this?

PRISCILLA: You have a nice baby. I'm sure Sally will take good care of him with Tucker. She told me you're going back to work.

(PRISCILLA *is interrupted by a noise from* TUCKER *in other room.*)

MAX: I'll go.

SALLY: Uncle Max has been helping with Tucker while Ashley . . .

PRISCILLA: Yes. (MAX *exits to the bedroom.*) Cam, honey, will you open up the bag? (CAMERON *does so.*) Get that little cap. He'll need it back home in the sun. He might as well get his first present now. (CAMERON *produces a little baseball cap with JESUS LOVES YOU on it.* PRISCILLA *puts it on* ASHLEY'S *head.* CAMERON *is about to close the case.*) Well, I guess we'll be going now.

HARRIET: Wait. (CAMERON *stops what he's doing. To* PRISCILLA.) Could I have a moment with you?

PRISCILLA: What for?

HARRIET: Just one moment. (*To* SALLY.) You mind?

SALLY: Of course not.

(SALLY *heads for the kitchen.* HARRIET *waits for* CAMERON *to also leave. He doesn't.*)

HARRIET: I don't want to involve you, Cam.

PRISCILLA: Why shouldn't he be here?

HARRIET: For the same reason Sally shouldn't be. I want to solve one problem at a time.

CAMERON (*to* PRISCILLA): It's okay, honey.

(CAMERON *goes into the kitchen.*)

HARRIET: We should get to know each other. Let's give it a try . . . the two of us . . . alone.

PRISCILLA (*dubious*): I don't know about that.

HARRIET: Look, you got to know Sally in Mississippi. She said things worked out pretty well.

PRISCILLA: She's Cam's sister.

Taboos | 101

HARRIET: That also makes her your sister-in-law.

PRISCILLA (*reluctantly*): I guess so.

HARRIET: Well, you guessed right. Which of course makes me the brother-in-law. But you never wanted to know Sally or me.

PRISCILLA: What you were doing was sinful.

HARRIET (*sarcastic*): Past tense? Whatever it is, we're still doing it. (*Beat.*) You didn't talk about Tucker with Sally?

PRISCILLA: Not really.

HARRIET: Before we know it, Tucker will be two. You didn't even come to his first birthday. Of course, Cameron did. (*Beat.*) You have no idea how important that was for Sally . . . to have Tucker's uncle here. And you're his aunt.

PRISCILLA: I just couldn't come.

HARRIET: Did Cam tell you about us?

PRISCILLA: Very little.

HARRIET: You weren't curious?

PRISCILLA: I didn't want to know . . . (*Beat.*) I was afraid. (*Beat.*) I'm still afraid.

HARRIET (*surprised*): Of us?

PRISCILLA: Your lifestyle.

HARRIET: What do you know about (*Sarcastic.*) our "lifestyle?" (*She waves her hand around the room.*) Does this look like a house of ill repute? (PRISCILLA *silently shakes her head.*) Well then?

PRISCILLA: It isn't what I thought.

HARRIET: I see. And if it hadn't been for Ashley's colic and your own problems you would still not have come?

PRISCILLA: I couldn't take his crying any longer. I thought I was losing my mind.

HARRIET: I know. (*Beat.*) Didn't you ever suspect why Cam brought Ashley here?

PRISCILLA (*she shakes her head*): I guess I was too upset.

HARRIET: Ashley needed my milk. It worked because Jan and Ashley are twins. You must have noticed the resemblance between them.

PRISCILLA (*stubbornly*): They're half-brothers. They share a father . . . nothing more. And they don't at all look alike!

HARRIET (*furiously*): Not alike? (*She rips off* ASHLEY's *baseball cap and hurls it across room.*) Just look at him without that creepy JESUS LOVES YOU cap. Look! (*She thrusts* JAN *practically in her face.*) Spitting images! (*Pause.*) I would like to see Ashley from time to time.

PRISCILLA: Out of the question!

HARRIET (*hurt*): Why?

PRISCILLA: It isn't right.

HARRIET: What isn't right?

PRISCILLA: I will not have my child exposed to a man-hating lesbian!

HARRIET: Who says I hate men? I don't happen to welcome them into my bed. But otherwise? Most of my patients are men . . .

PRISCILLA: We're talking about family . . . not your office.

HARRIET: Okay . . . let's talk family. I love my father . . . I adore my brother . . .

PRISCILLA: Your character then. My child will have a man to look up to.

HARRIET: All children need male role models . . . and male love . . . and male bonding within a family.

PRISCILLA: But . . .

HARRIET: But what? Tucker and Jan have two mothers and two uncles. At least one of them spends a lot of time with them.

PRISCILLA: Uncles and fathers are not the same.

HARRIET: I didn't say they were. Some biological fathers can't even be uncles . . . let alone fathers . . . and some uncles are almost as good as fathers. Just take Max in there . . . with Tucker. At least in our family . . . Sally's and mine . . . we can choose the right male for our sons.

PRISCILLA: No.

HARRIET: Why not?

PRISCILLA: Because it isn't right!

HARRIET: Says who?

PRISCILLA: It isn't natural!

HARRIET: It isn't natural, so it isn't right? You didn't become a mother the natural way. Does that make it wrong? Is your family now missing a mother?

PRISCILLA: No. No. No! You are wrong!

(*Impasse.* HARRIET *tries again.*)

HARRIET: They'll always have a special bond, Jan and Ashley. They are twins. Are you going to deny them that?

PRISCILLA: They're not twins! Yours is much fatter.

HARRIET: You know damn well why that is so . . . with Ashley born prematurely and then his colic. But they're twins . . . one fatter and the other thinner!

PRISCILLA (*stubbornly*): All right . . . biological twins. But you are talking about enzymes . . .

HARRIET (*now truly irritated*): Oh, for God's sake! What do you know about enzymes?

PRISCILLA: I know!

HARRIET: You know what, exactly? What? Give me some interesting facts about enzymes! Go on! Surprise me!

PRISCILLA: I looked it up on the Internet.

HARRIET (*explodes*): The Internet? Jesus Christ!

PRISCILLA: Don't take our Lord's name in vain!

HARRIET (*about to lose her temper completely*): Jesus! Half-brothers! And what exactly was my role in all this? To provide the right enzymes to cure his colic . . . and nothing else?

PRISCILLA: It was your embryo . . . but my baby.

HARRIET (*sardonic*): Why don't you go back to the Internet to see what you can find under "genes"?

PRISCILLA: I don't need the Internet to tell me about my child. For nine months, Ashley and I formed a relationship. You understand? A relationship that'll last until death! I *talked* to Ashley while he was in me . . . when you didn't even know what had happened to your embryo! Sure, I'll always be thankful to you for that gift . . . but I'm not just an incubator for your embryo. Ashley is *my* child. He's been baptized. (*Almost hysterical.*) *Baptized!* Do you hear that? This child was born again when he was ten days old!

HARRIET (*now calmer*): I'm not going to argue with you about nurture versus nature. Especially not when I nursed that kid for the past month. Or do you think I'm just a milk cow? (*A long, seething pause.* CAMERON *and* SALLY, *having heard the argument, come quietly out of the kitchen and look on.*) You know, while we're talking about twins . . .

PRISCILLA: Half-brothers.

HARRIET: Whatever. I hope you don't think you have any rights to my child?

PRISCILLA: To Jan? No . . . I don't have any.

HARRIET: Well that's a consolation.

PRISCILLA: But Cameron's the father, so you'd need to check with him.

HARRIET (*explodes*): I see. You think I have no rights with respect to Ashley as the egg donor, but your Cameron's puny little sperm will give him rights to my son?

SALLY: Harriet, please . . .

PRISCILLA (*lowers her head and starts praying*): My Lord and Savior. I thank you for having restored my equanimity. But now turn your bountiful gaze upon this sinner Harriet that she may repent . . .

HARRIET: Jesus Christ!

(MAX *has come out of the bedroom.*)

MAX: There's a two-year-old boy in this bedroom who can hear every word you're saying. A boy to whom I've just read this book! (*He waves the book.*) You remember . . . "It doesn't matter how many mommies or how many daddies your family has . . . It doesn't matter if your family has sisters or brothers or cousins or grandmothers or grandfathers or uncles or aunts . . . Each family is special. The most important thing about a family is that all the people in it love each other." Tucker doesn't understand any of that yet, but he soon will. And eventually, so will the other two boys. Aren't you two ashamed of yourselves? If you have something to resolve, why don't you discuss it like sensible adults?

HARRIET (*after a beat*): I'm sorry.

(*A long pause.* MAX *notices something hanging out of the suitcase.*)

MAX (*to* PRISCILLA): Is that a doll?

(*Somewhat surprised,* PRISCILLA *nods.* MAX *goes over to the suitcase, lifts the lid to replace the doll inside, looks inside the case.*)

That's quite a collection you've got here. How come you brought them?

PRISCILLA: They were mine as a child.

HARRIET (*a touch of sarcasm*): You always travel with your dollies?

PRISCILLA: I brought them for Ashley . . . for company.

HARRIET (*even more sarcastic*): Don't you think Ashley needs male company rather than girlie dollies?

PRISCILLA: Some of them are boy dolls.

MAX (*kind tone*): Which is your favorite?

(PRISCILLA, *unsure of what else to do, goes over to the case and brings out the battered old doll we saw her with before.*)

How long have you had her?

PRISCILLA: Since I was four.

MAX: She's lasted well. You must have given her a lot of love over the years. (*He addresses the others.*) Listen. I've got an idea. Why don't you people pull up some chairs? And put the children into the crib . . . it may be their last time together for a long time.

(*Baffled, but going along with it,* SALLY *and* HARRIET *move toward chairs, but* PRISCILLA *and* CAMERON *remain standing.* MAX *addresses them.*)

Come on . . . why don't you give it a try. What have you got to lose?

(PRISCILLA *and* CAMERON *start to sit down, but on opposite sides of the table.*)

Not that way. Two of you on one side . . . and two on the other. And Priscilla, mind if I borrow some of these dolls? I think it might help. I promise to be gentle with them.

PRISCILLA (*hesitatingly*): I guess it's okay.

MAX: Thank you.

(PRISCILLA *and* CAMERON *place themselves close together on one narrow side of the rectangular table and* HARRIET *and* SALLY *on the opposite side.* MAX *sits down in the center of one of the unoccupied sides of the table. He addresses the entire group as he looks around, from time to time favoring one group over another.*)

You realize, of course, that I don't usually do this sort of thing. My business isn't to resolve problems. I'm not a judge . . . I'm usually an advocate for the accused.

(*He notices* PRISCILLA *whispering to* CAMERON. *He addresses* PRISCILLA.)

But here we have two accusers who are also both defendants. Why don't you let me take both your sides. (*To* PRISCILLA.) I barely know you but my fellow sperm donor over there (*He points to* CAMERON.) can vouch for me.

PRISCILLA (*hesitates*): I don't know . . .

CAMERON (*quickly interrupts*): Prissy, let him try.

MAX: Okay?

PRISCILLA: I guess so.

MAX: Here (*Pointing to* SALLY *and* HARRIET.) we have two highly intelligent professionals . . . too busy in their work to have much time for random encounters . . . who wanted (*He draws quotation marks in the air with his fingers.*) "to get married" after they were lucky enough to have met each other.

PRISCILLA: Two women can't get married!

MAX (*calmly*): Priscilla! I drew quotation marks around those words. The fact is . . . Harriet and Sally wanted (*He again draws quotation marks in the air with his fingers.*) "to get married" . . . for better or for worse, until death do them part.

PRISCILLA: But . . .

MAX (*interrupts her, raising his hand*): Both of them now have children . . . which is not that unusual . . .

PRISCILLA: It's pretty unusual back home.

MAX: I can assure you, it's anything *but* in San Francisco. However, what is unusual is the familial relationships of the mothers to the sperm donors. And especially the further relationship of one of them (*He points first to* CAMERON.) to a woman . . . meaning you, Priscilla . . . who became (*He again draws quotation marks in the air with his fingers.*) "a mother" by decidedly unnatural means.

PRISCILLA: You can't put quotation marks around that word! I *am* a mother!

MAX: Of course you're a mother. The quotation marks referred only to the means whereby you became one. And now, let's get to the heart of this mess!

SALLY: Max! A mess? If you think that giving birth to two wonderful children . . .

HARRIET: Three children! The eggs were all fertilized in San Francisco!

MAX: If I remember right, "mess" means something like "untidy, disordered, or unpleasant." And doesn't that pretty much sum up where we are all right now? Eight persons . . . two brothers to two sisters . . . two sperm donors to three sons . . . two uncles to three nephews . . . and a married couple with a marriage certificate and another couple without one. Anyway, let's call this part of the table (*He points to the side where* CAMERON *and* PRISCILLA *sit.*) "Mississippi." And here (*Now he points to the side where* SALLY *and* HARRIET *sit.*) is "San Francisco." We need some distance between you . . . as in real life. Remember, you don't live in the same house.

> (MAX *reaches into the suitcase and brings out three dolls:* PRISCILLA's *favorite and two smaller ones. He addresses* SALLY *while grouping the three dolls close together in the middle of the table.*)
> Sally, I need three sheets of paper and a pen. . . . and some tape.
> (SALLY *brings them to him, whereupon he writes each child's name in big letters on the sheets while talking.*)
> This is Jan . . . this one can be Tucker . . . and this one, Ashley. And to avoid any confusion, let me label them.
> (*He tapes the names to the front of each respective doll.*)
> What I would now like each of you to do in succession is to move each child to the adult couple where you think they belong. Understood? (*They all nod.*) Now let's start with you, Priscilla.
> (PRISCILLA *quickly reaches for the* ASHLEY *doll and moves it right next to her.*)

PRISCILLA: There you are.
MAX: Don't you want to move the other two?
PRISCILLA: No . . . I have made my point.
MAX (*to* CAMERON): Cam.

> (CAMERON *leans over to move the* TUCKER *doll all the way across the table next to* SALLY. HARRIET *quickly reaches over and moves that doll so it is between her and* SALLY.)
> Harriet . . . it's not your turn.
> (MAX *reaches over and moves* TUCKER *back to the position next to* SALLY *where* CAMERON *had placed it originally. Then he turns to* CAMERON.)

That's all?

CAMERON: Could I pass until the end?

MAX: Sure. (*To* SALLY.) In that case, you're next, Sally.

(SALLY *quickly carries out the* TUCKER *move that* HARRIET *had initiated and then moves* JAN *right next to him, thus placing both* TUCKER *and* JAN *right between* SALLY *and* HARRIET.)

No other move, Sally?

SALLY (*hesitates, then moves* ASHLEY *away from* PRISCILLA *slightly toward the center.*) That's it.

MAX (*to* HARRIET): Harriet?

(HARRIET *reaches across table and moves the* ASHLEY *doll all the way to the center.* PRISCILLA *immediately reaches over to put him back next to her. To* PRISCILLA.)

Priscilla . . . you can't do that! It's not your turn.

PRISCILLA: Ashley belongs to me!

MAX: Priscilla! These are dolls, not children.

(MAX *moves the* ASHLEY *doll back toward the middle, where* HARRIET *had placed it.*)

PRISCILLA: I don't care. Ashley belongs here.

(PRISCILLA *moves* ASHLEY *back.* HARRIET *reaches all the way over the table—practically lying on it—for the* ASHLEY *doll, but before she can move it,* PRISCILLA *grabs it as well.*)

HARRIET (*to* CAMERON): Your wife seems to think children are dolls.

CAMERON (*to* HARRIET): That was unfair. (*Gently to* PRISCILLA.) Prissy . . . let go. It's just a game.

(*Reluctantly,* PRISCILLA *relinquishes her grip, whereupon* HARRIET *moves the* ASHLEY *doll back to the center.*)

MAX (*to* HARRIET): Any other moves you wish to make.

HARRIET: No.

MAX (*to* CAMERON): You passed earlier. You have two more moves.

(CAMERON *hesitates for a moment, then reaches for the* ASHLEY *doll and moves it still closer to "Mississippi," but not as close as* PRISCILLA *had placed it.*)

Okay. In that case . . .

CAMERON: I'm not finished yet, Max.
>   (CAMERON *reaches way over the table for the* JAN *doll and moves it from* HARRIET *and* SALLY *toward the middle, though still closer to the "San Francisco" than the "Mississippi" position.*)
>   That's it.

HARRIET (*under her breath*): Jesus!

MAX (*after looking slowly around the table*): We're getting close to the end. You can each make one more move . . . but only *one!* Let's do it in the same order as before. Priscilla, you first.
>   (PRISCILLA *moves the* ASHLEY *doll back next to her;* CAMERON *gestures that he does not wish to do anything;* SALLY *does likewise; whereupon* HARRIET *looks up angrily. She reaches over to* ASHLEY—*again practically lying on the table*—*ready to move the doll back toward the middle when* MAX *interrupts . . .*)
>   Harriet, remember, *one* move only.
>   (HARRIET *drops her hand, sits back in her chair, and then quickly grabs the* JAN *doll and moves it back next to her from the position where* CAMERON *had placed it last. Long pause.*)

HARRIET (*to* PRISCILLA): If this (*She points to the* ASHLEY *doll.*) had reflected reality, who would've breast-fed Ashley the last four weeks?

PRISCILLA: May the good Lord forgive you for such cruel words.

HARRIET: You think it's cruel asking why you won't move Ashley partly to our side?

PRISCILLA: Yes! (HARRIET *suddenly makes a play for the* ASHLEY *doll. In reaction,* PRISCILLA *grabs at it. In the ensuing tug of war, the doll breaks in half.*) You're wicked! Look what you've done!
>   (PRISCILLA, *upset that her favorite doll has broken in two, holds the pieces in her hands.*)

CAMERON (*to* MAX): Some game!

MAX: This wasn't a game. (*Beat.*) Cam . . . Harriet was wrong thinking she could just let you use an embryo as you'd see fit, and now she knows it. When she put Ashley to her breast . . . she discovered a mystery she was not prepared for. You can't blame her for feeling Ashley belongs to her, too.

PRISCILLA (*to* HARRIET): You're a doctor . . . and you handled the embryo as a doctor. I converted it into a baby. I already told you: I'm the mother.
CAMERON: What about us, Max? You and me? Are fathers just sperm donors?
MAX: When Sally gave birth to Tucker . . . how did you feel? A little envious, perhaps?
CAMERON: More than a little. And I asked the Lord to free me of that envy.
SALLY: Envious of me . . . your sister?
CAMERON: Of Max. He had the advantage of being close to you two and Tucker.
MAX: You should have come more often. (*Beat.*) But it isn't that simple, is it? This table top . . . it's a map of your emotions. The messiness is not your conflicting rights . . . it's the emotional bonds you have to so many different people here. Even you, Priscilla, will discover that you aren't just a mother. You're also an aunt. Let me remind you that in a court the only thing that matters is the welfare of the children. We all know I'm no judge, but perhaps I can be an arbitrator . . . one who focuses on the children. Let me drop a little bomb among you to see what happens. (*Beat.*) I think our situation is crying out for a kibbutz-like solution. (*Beat.*) Not that it works everywhere . . . but it might here. In an Israeli kibbutz, children—especially small ones—are brought up together in the community but with plenty of different bonding opportunities with each other and with their parents.
SALLY: Get real, Max. You want us to move to Israel?
MAX: Of course not. In San Francisco . . . parents like you and Harriet are accepted. I know it's a lot to ask, but what if Priscilla and Cameron moved here? We also pay taxes here and need CPAs . . . so what if you all lived close by . . . if the three boys were brought up together during the day?
  (MAX *puts all three boy dolls together but about equidistant from* SALLY-HARRIET *and* CAMERON-PRISCILLA.)
SALLY (*quickly interjects*): But live in their own homes at night and weekends!
HARRIET: And there's total separation of church and state . . .

CAMERON: But only until they are old enough to go to school . . .

SALLY: Why call it a kibbutz? You're actually suggesting a nursery, a child-care center, run by . . .

MAX: Me! It's what we've been doing for the past few days, anyway. This is better than the usual nursery or even a kibbutz, because the children are related and it's not being run by a stranger.

>   (MAX *slowly picks up the three child figures and places them around himself near the middle of the table.* HARRIET, SALLY, *and* CAMERON *suddenly stare at* MAX, *surprised and relieved.*)

SALLY: You're willing to do that?

HARRIET: What about your job?

MAX: I might as well confess something to all of you. I love kids, Tucker has shown me that. And as my sister has told me more than once—I don't just flirt with an issue . . .

HARRIET (*interrupts*): He marries it!

MAX: That's more or less it.

SALLY (*startled*): What exactly do you want to marry?

MAX (*slight smile*): Well . . . not exactly "marry," because that word is being bandied about a bit too much around here . . . and it's making everyone nervous. But I want to see what it's like to be a house husband.

PRISCILLA (*shocked*): What kind of a husband?

MAX: The kind that's willing to stay around at home . . .

PRISCILLA: While the wife works?

MAX: What alternative is there . . . unless one of them has an independent income?

PRISCILLA (*taken aback*): You're aiming to do that for the rest of your life?

MAX (*now openly smiling*): Hardly that. But while the kids are preschool? Why not? A lawyer can take off for a few years and then return to his profession.

SALLY (*slightly suspicious*): You mean you just want to run our kibbutz to get some practice with young kids?

MAX (*still slightly bantering*): You don't think that's enough of a reason?

SALLY: Well . . . it *could* be . . . but I have a feeling there's more to it. (*Beat.*) Is there?

MAX (*now serious*): I guess there is.

SALLY: Well then, let's hear it!

MAX: Harriet has discovered that trying to be an egg donor is not enough. Not if you then get to see the baby and nurse it.

SALLY: Stop, Max! Stop right there! (*She reaches over for the* TUCKER *doll and clutches it to her chest.*) Are you about to tell me that you had some sperm donor epiphany about Tucker?

MAX: Not an epiphany . . .

CAMERON: I know exactly what Max is saying . . .

SALLY (*angrily interrupting*): Cam! You keep out of this . . . it's none of your business.

CAMERON: I'm Tucker's uncle . . .

SALLY: Jesus Christ!

PRISCILLA: Don't take the Lord's name . . .

HARRIET: Shut up! All of you! Max, what exactly are you saying?

MAX: Sally and Harriet! You know me well enough that I won't renege on our understanding . . .

SALLY: In other words, a sperm donor and an uncle, but nothing more?

MAX: As far as Tucker is concerned? Yes, that's all. But I've now realized that the next time, I don't just want to be a sperm donor, but a full father.

CAMERON: Praise the Lord! He also wants to do it the natural way.

MAX: Listen, Cam, I'm not talking about fertilization. I don't care whether it's in bed or under the microscope. I'm talking about what happens after fertilization . . . actually being a *father*.

SALLY: So you're volunteering to run this Californian kibbutz as a way to discover how good you are at full-time fathering?

MAX: Yes. I'm willing to take a year off from my job if you four are willing to pay me to run this kibbutz during the day. And then we'll see.

HARRIET: I'm willing to give it a try.

SALLY: Me, too. (*She moves the* TUCKER *doll back to the center. To* CAMERON.) You can be a father as well as a double-uncle. And Priscilla, you might even learn to be an aunt.

CAMERON (*to* PRISCILLA): Prissy . . . ?

PRISCILLA: "Prissy" what?

(CAMERON *falls silent.* PRISCILLA *picks up her three dolls and returns them to the suitcase.*)

MAX: You know, the kibbutz movement started in the Holy Land. Surely starting one in San Francisco can't be that sinful?
   (PRISCILLA *stops packing away the dolls, almost as though she is seriously thinking about the offer.*)
PRISCILLA (*to* CAMERON, *pointing to the corner of the table.*) Let's pray.
   (CAMERON, *not without embarrassment, goes over to* PRISCILLA *and kneels down beside her.* SALLY *and* HARRIET *withdraw to one side to give the praying couple privacy.* MAX, *however, listens carefully.*)
   Heavenly Father. Since the kibbutzes . . .
MAX (*he interrupts in low voice*): I don't think there's such a plural. Maybe it's like fish or sheep.
PRISCILLA (*starts over*): Heavenly Father. Since kibbutz were started in the Holy Land, surely starting one in San Francisco may not be sinful.
HARRIET (*sotto voce*): Thank God!
PRISCILLA (*looks at* HARRIET *briefly before continuing*): But they're not for Christians.
HARRIET (*louder, not being able to contain her shock*): Jesus!
PRISCILLA: Give us the strength not to compromise as we bring up Ashley within the Christian faith, even though he was born by artificial . . .
CAMERON (*interrupts*): Alternative . . .
PRISCILLA: I mean alternative means. And since we still have four embryos . . .
CAMERON (*interrupts*): Five embryos . . .
PRISCILLA: Are you sure?
CAMERON: Sure.
PRISCILLA: Praise the Lord! Even better! O Lord . . . since you taught us to go forth and multiply . . . surely it is less sinful to use them . . .
CAMERON (*interrupts*): . . . than to destroy them. So bless us with siblings to Ashley . . .
PRISCILLA: Which we shall bring up as thy faithful servants in a true Christian household. Amen.
CAMERON: Amen.
HARRIET: Wait a moment! Those are my eggs!
SALLY: You've got to ask her.
MAX: If Ashley is raised in the kibbutz, they all have to be. And you heard Harriet about the separation of church and state.

(PRISCILLA *gets up and takes* ASHLEY. *She turns to* CAMERON.)
PRISCILLA: Get the suitcase, honey.
SALLY: You're not going now?
PRISCILLA: Cam, get the case.
MAX: I take it that means you don't approve of the kibbutz option?
PRISCILLA: The suitcase, please, Cam.
CAMERON: Honey, I don't think we should leave just yet.
PRISCILLA (*bitter*): Fine. You stay. But I'm taking my child to Mississippi.
MAX: Priscilla, I don't think you should leave. How do you know the Lord heard only you?
(PRISCILLA *goes toward the door.*)
CAMERON: Prissy! You said "a true Christian household." But the Bible says, "By wisdom a house is built, and through understanding it is established."
PRISCILLA (*low voice*): Proverbs 24:3. (*Dismissive.*) That's the Old Testament!
CAMERON: Well then remember what Jesus says in John 14, verse 2. (PRISCILLA *pauses.*) "In my Father's house are many mansions: if it were not so, I would have told you."
(PRISCILLA *turns to face us, a startled expression on her face, as the lights fade.*)
(*End of scene 12. End of play.*)